Practical Road Safety Auditing

TMS Consultancy

Published by Thomas Telford Publishing, Thomas Telford Ltd, 1 Heron Quay, London E14 4JD.
URL: http://www.thomastelford.com

Distributors for Thomas Telford books are
USA: ASCE Press, 1801 Alexander Bell Drive, Reston, VA 20191-4400, USA
Japan: Maruzen Co. Ltd, Book Department, 3–10 Nihonbashi 2-chome, Chuo-ku, Tokyo 103
Australia: DA Books and Journals, 648 Whitehorse Road, Mitcham 3132, Victoria

First published 2001

A catalogue record for this book is available from the British Library

ISBN: 0 7277 2938 1

Typeset by Academic + Technical Typesetting, Bristol

Printed and bound in Great Britain by MPG Books, Bodmin

Contents

Foreword

In the UK, we have an enviable reputation for improving road safety. Today there are fewer road deaths than in 1930 while the number of vehicles on the roads has risen from 2.3 million to over 27 million.

More recently there have been further reductions in casualties resulting from road accidents. Since 1987, road deaths and serious injury have fallen by around 40% and Britain is recognised as having amongst the safest roads in the world. However, reductions in walking and cycling have contributed to casualty reductions for pedestrians and cyclists, and Britain has a poor record for child pedestrian safety when compared with other European countries.

There is therefore no room for complacency in road safety, and the government has set new casualty reduction targets for 2010. These include a 40% reduction in all fatal and serious casualties and a 50% reduction in fatal and serious casualties involving children. Road safety improvements will only be achieved if all relevant agencies work together to adopt these targets and put realistic plans in place to meet them.

A key part of a comprehensive road safety plan must be the safe design of our road infrastructure. Road Safety Audits provide a vital role in checking that roads have indeed been designed and built to the highest safety standards. This book is a step forward in improving the quality of Safety Audits as it brings together best practice in uncovering and providing solutions to safety problems from some of the most expert and longest serving Safety Audit practitioners in the country.

Preface

For many years we have believed that Road Safety Audit has had the potential to make a significant contribution to highway safety. The formal audit process involves looking at schemes throughout the various stages of design, and trying to identify road safety problems. The Auditor then goes on to recommend solutions to the problems that have been described.

The benefits of Road Safety Audit rely on the Safety Auditor being able to identify genuine 'safety problems' – features within the design that if left unaltered would actually lead to accidents. The Auditor must also be able to recommend solutions that will work – based on experience of similar treatments elsewhere.

Many organisations – the Highways Agency, the Institution of Highways and Transportation (IHT), local authorities, and national government agencies overseas – have written detailed Road Safety Audit procedures guiding staff through the process of Safety Audit. Very little has been written down about the safety problems and solutions themselves, at least not in a format easily accessible to Safety Auditors. Information is available in many disparate forms – Highways Agency Advice Notes and Standards, Transport Research Laboratory (TRL) reports, local authority reports, road safety plans and individual case studies.

This book attempts to bring some of this information together in a format that the Safety Auditor can use at a practical level. The book contains many examples of real road safety problems backed up by accident statistics, together with examples of successful solutions, again backed up by comprehensive evidence.

The book has been written by practising Safety Auditors, who between them have carried out over 1000 audits. The aim of the book will be to teach and to inform – a guide for those commissioning Road Safety Audits, those carrying out Road Safety Audits, and for those whose schemes are being audited.

Acknowledgements

This book is dedicated to all those Safety Auditors across the world who have turned up for a pre-arranged 'post opening' Stage 3 Audit only to find that the wearing course has not yet been laid, that it's pouring with rain, and that there are no bacon sandwiches left in the resident engineer's portacabin.

The authors would like to acknowledge the assistance of all of the staff at TMS Consultancy in the preparation of this book, and in particular Gudbjorg Lilja Erlendsdottir for all her hard work. The assistance of Rod Chaplin (formerly with the Highways Agency) is acknowledged in providing data used in Chapter 5. The authors are also grateful to Ian Appleton from Transfund New Zealand, Frank Navin from the University of British Colombia in Canada, and Phil Jordan from VicRoads in Australia for their contributions to the International section of the book. The assistance of Paul Forman and his Investigations and Risk Management Team at TRL in looking over Chapter 7, is also gratefully acknowledged.

The authors are Directors of TMS Consultancy, an independent traffic and road safety consultancy firm, established in 1990, and based in Coventry, UK. TMS has carried out more than 1800 Road Safety Audits since 1991, and has worked throughout the UK, and internationally. In addition to carrying out consulting work, TMS runs Road Safety Audit and other training courses.

The views expressed are those of the authors alone. TMS Consultancy and Thomas Telford do not accept responsibility for the actions of any person or organisation that claims to be acting as a result of advice taken from this publication.

A number of photographs are included in this book. The inclusion of any photograph should not be seen as a direct criticism of any individual highway authority or design organisation. The purpose of the photographs is to illustrate largely generic issues, many of which can be seen throughout the UK. The authors are grateful to TMS staff and others for providing photos for inclusion in this book.

Steve Proctor
Martin Belcher
Phil Cook

Directors, TMS Consultancy
Barclays Venture Centre
University of Warwick Science Park
Sir William Lyons Road
Coventry CV4 7EZ
024 7669 0900
info@tmsconsultancy.co.uk
www.tmsconsultancy.co.uk

1. Introduction

Definition of Road Safety Audit

Road Safety Audit is a systematic process for checking the *safety* of new schemes on roads. It should be based on sound safety principles and should ensure that all highway schemes operate as safely as is practicable by minimising future accident numbers and severity.

Safety Audit provides road safety engineers with an opportunity to feed their experience into the highway design process. Road Safety Audit should improve the awareness of safe design practices by all concerned in the design, construction and maintenance of roads.

Safety Audit considers the safety of all road users and in particular vulnerable road users such as the visually and mobility impaired, cyclists, pedestrians, equestrians, motorcyclists, children and the elderly. Safety Audit is not an opportunity to redesign a scheme or to make changes to design with no apparent link to a safety issue. It is not intended to be a technical check on the design elements nor a design standards check. These should be carried out independently of the Safety Audit.

Although Safety Audit does look at scheme design from the road user's point of view, it is not in fact a 'road user audit', which aims to ensure that each road user has been adequately catered for within a scheme. For example, a Safety Audit of a roundabout design should be concerned about the potential for cycle accidents, particularly if it is intended that cyclists should use the circulatory carriageway. A road user audit, on the other hand, should be concerned to ensure that there is provision for cyclists within the design.

History

British traffic engineers first developed the idea of a Road Safety Audit as a safety check for new and improved road schemes in the early 1980s.

However, the concept of a safety checking procedure within transportation had existed for over 150 years.

A study of the development of the British railway network from the 1830s shows how Railway Engineers and their regulators grasped the concept of Safety Audit as the key to a safe transport system. In the early days of this form of mass transportation, accidents were common as the pioneers came to grips with the demands of moving people and goods, at previously unknown speeds, around the country.

As the railway industry developed, the British Government, through the Board of Trade, appointed senior army officers from the Royal Engineers to investigate the frequent accidents that were taking place. From this work they made recommendations to stop similar occurrences taking place on both existing and new lines. Before each new railway line was opened, or modifications were carried out, these officers were charged with carrying out an inspection of all the operational safety aspects, and unless the project obtained a clean bill of health the trains did not run.

Figure 1.1
Road Safety Audit has its origins in the railway industry

The Road Safety Audit process in the UK started to gather pace when safety engineers realised that they were carrying out accident remedial schemes on relatively new roads. Adopting the principle of 'prevention is better than cure', they decided to use some of the safety experience they had gained from the remedial work, and design safety into new road schemes. The Institution of Highways and Transportation Guidelines on Accident Investigation and Prevention produced during this time included an emphasis on 'safety checking', as an accident prevention mechanism.

Since then, the important milestones in the development of Safety Audit in the UK have been the following.

- The Road Traffic Act 1988[1] which states that 'in constructing new roads, (local authorities) must take such measures as appear to the authority to be appropriate to reduce the possibilities of such accidents when the roads come into use'. The wording was, in fact, very similar in the 1974 Act. This is sometimes interpreted as a statutory basis for Road Safety Audit work.
- In 1990 the Design Standard HD 19/90 and Advice Note HA 42/90 'Road Safety Audits'[2,3] were introduced as part of the Design Manual for Roads and Bridges (DMRB). This made Safety Audit mandatory on trunk roads and motorway schemes from 1991 onwards.
- In 1990 the IHT produced guidelines 'The Safety Audit of Highways'.[4] These were adopted by many local authorities who started to carry out Safety Audits on local roads.
- In 1994 the Design Standard and Advice Note were revised: HA 42/94, HD 19/94 'Road Safety Audits'.[5,6]
- In 1996 the IHT Safety Audit guidelines were revised.[7]
- In 2000 the Highways Agency commenced a thorough review of UK Safety Audit practice with a view to a radical overhaul of the Standard and Advice Note.

By the end of the twentieth century, most local Highway Authorities in the UK were undertaking Road Safety Audits on road schemes within their areas of responsibility.

Design standards and safety

Despite the use of current design standards, accidents do occur on some schemes when completed. There are a number of reasons for this.

On local roads, particularly in urban areas, it is not always possible to keep to design standards, while at the same time constructing schemes within highway land and within budget. One of the benefits of Road Safety Audit is that the auditor can suggest measures that mitigate against the constraints imposed upon the design.

Combinations of features individually designed to standard can cause problems, for example, minimum standards of horizontal and vertical alignment which may lead to visual deception.

'Real world' accidents are not always covered by standards. For example, the standard on safety fences recommends that impact protection should be installed if an embankment exceeds six metres high. Many accidents involving serious injury occur when vehicles descend embankments less than this height.

Figure 1.2
Vehicle fire following loss of control down embankment

Finally, the definition of safety often understood within the standards relates to how engineers design roads as opposed to how individuals use them. Visibility splays, size and location of signs, and protection of street furniture all relate to the design speed on a new road. But if road users perceive the road to be faster, they will drive it as such, sometimes leading to accident occurrence.

Requests from Safety Auditors that go beyond standards can lead to conflict with designers, and arbitration may be required. This is dealt with in Chapter 3.

International development of Safety Audit

Formal Safety Audit procedures have been developed in a number of countries, following the initiatives taken in the UK.

During the early 1990s, work was carried out in Australia, Denmark and New Zealand. Since then, national and local governments in Canada, France, Greece, Hong Kong, Iceland, Ireland, Italy, Malaysia, the Netherlands, Peru, Singapore, and the United States have been investigating the development of Road Safety Audit.

Chapter 6 of this book contains more information on international Safety Audit procedures.

Figure 1.3
Examples of International Road Safety Audit procedures

References

1. HMSO. Road Traffic Acts 1974 and 1988
2. HA 42/90 (1990). Department of Transport
3. HD 19/90 (1990). Department of Transport
4. IHT (1990). Guidelines for the Safety Audit of Highways
5. HA 42/94 (1994). Highways Agency
6. HD 19/94 (1994). Highways Agency
7. IHT (1996). Guidelines for the Safety Audit of Highways

2. Safety Audit procedures

Trunk road standards

In 1990, the Department of Transport published an Advice Note and Standard on Road Safety Audits as part of the Design Manual for Roads and Bridges (DMRB).[1] The DMRB is the prime source of guidance for trunk roads and motorways in the UK and is used as a 'best practice' document for local roads.

The Standard (HD 19/90) made Road Safety Audits mandatory on 'all schemes on trunk roads including motorways which involve permanent changes to the existing highway layout' from April 1991. The publication of the Standard resulted in many local authorities beginning to carry out Safety Audits on local roads as well as on the trunk roads and motorways for which they were agents.

In 1994, the Highways Agency produced revised versions of the Advice Note and Standard.[2] The new Standard (HD 19/94) describes the various stages of audit, the make-up and appointment of the Safety Audit Team, the audit brief and reporting procedures. It also describes how road accidents should be monitored after schemes have been implemented.

The Advice Note (HA 42/94) elaborates on the elements described in the Standard and provides an illustrative report showing a problem/ recommendation format for report writing.

On trunk road and motorway schemes, Safety Audit should be carried out at three formal stages: Preliminary Design (Stage 1), Detailed Design (Stage 2) and Pre-Opening (Stage 3).

IHT Guidelines

In response to the growing number of local authorities that were starting to carry out Safety Audits, the Institution of Highways and Transportation (IHT) produced 'Guidelines for the Safety Audit of Highways' in 1990.[3]

Figure 2.1
The IHT Road Safety Audit Guidelines

The IHT revised the Guidelines in 1996.[4] The revision was based on best practice gleaned from local authorities and consultancies that had been carrying out audits for some years.

The Guidelines describe how safety practice has developed since the 1980s, outline safety principles involved and describe procedures. The Guidelines also set out a code of good practice and provide some sample checklists that can be used by Safety Auditors. Legal implications are discussed together with the implications of the Health and Safety at Work Act 1974 and the Construction (Design and Management) Regulations 1994.

One of the major differences between the IHT Guidelines and the Standard is the recommendation that an audit should be carried out at the 'feasibility' stage (Stage F). IHT considers that if a scheme is audited early in the design process any fundamental safety problems can be addressed.

Local authority procedures

Many local authorities use the Advice Note and Standard together with the IHT Guidelines as a basis for carrying out their Road Safety Audits. However, some authorities have developed their own Safety Audit procedures to reflect local conditions more specifically geared to their own particular needs. Hampshire and Oxfordshire County Councils are examples of local authorities that have developed their own procedures.[5,6]

Local procedures are particularly important in the current climate when local authorities have a variety of client/consultancy splits within engineering departments. Some local authorities carry out their own highway design work while others use external consultants. In addition, some local authorities use external consultants to carry out the Safety Audit task.

Another important issue is the way local authorities audit development-led schemes. Some authorities audit all schemes related to new developments on an in-house basis while others insist that an independent Safety Audit be carried out for the developer.

Local procedures generally cover the following items:

- what schemes are audited;
- what information is provided for the Safety Auditors;
- who carries out the Safety Audits;
- how Safety Auditors are trained;
- how Safety Audits are to be reported;
- who writes exception reports;
- who arbitrates in the event of unresolved issues;
- how development-led schemes are audited.

In 1995, IHT carried out some research among Safety Audit practitioners to produce a 'Review of Road Safety Audit Procedures'.[7] IHT analysed completed questionnaires from 47 local authorities and 15 consultancies. At the time, 17 local authorities and four consultancies used their own Safety Audit procedures and guidelines (generally based on either IHT Guidelines or the Advice Note and Standard). It is likely that more local authorities have now drawn up Safety Audit procedures.

References

1. Department of Transport (1990). 'Design Manual for Roads and Bridges'
2. Highways Agency (1994). HA 42/94, HD 19/94
3. IHT (1990). 'Guidelines for the Safety Audit of Highways'
4. IHT (1996). 'Guidelines for the Safety Audit of Highways'
5. (1996). 'Hampshire County Council Safety Audit Procedures'
6. (1998). 'Oxfordshire County Council Safety Audit Procedures'
7. IHT (1995). 'Review of Road Safety Audit Procedures'

3. The Safety Audit process

How to carry out a Safety Audit

A Safety Audit tries to identify potential road safety problems and suggests ways in which these identified problems can be minimised. There is no set way to undertake Safety Audits and individual Safety Auditors will no doubt have their own methods. There are however fundamental rules that should be applied when carrying out Safety Audits.

The Safety Audit should be carried out by a team consisting of at least two people

Experience has shown that two people carrying out a Safety Audit will identify more potential safety issues than a single Safety Auditor.
In many cases the senior Auditor will be the Team leader with the second person being Team member.

Figure 3.1
Two Auditors on site in inclement weather conditions

The number of items to be checked, particularly on large highway schemes, can be onerous. Large numbers of detailed scheme plans may have to be examined. In addition, it is helpful to be able to discuss possible recommendations with another Safety Auditor. Using a Safety Audit Team helps to avoid the situation in which a single Safety

Auditor can sometimes overemphasise a particular issue because of that individual's experience or even prejudice.

There must be substantial road safety engineering experience within the Safety Audit Team

Without road safety engineering experience and an understanding of safety principles, a Safety Audit can become simply a design check. Road safety experience is essential in enabling Safety Auditors to identify road safety problems and assisting them in making appropriate recommendations. Chapters 4 and 5 provide extensive information on the subject. In addition to experience, Safety Audit Team members should have received adequate training from an approved training provider. This issue is discussed in more detail in Chapter 8.

Figure 3.2
Delegates working on a recent RoSPA Road Safety Engineering course

A site visit must be carried out as part of the Safety Audit

It is very important that a comprehensive site visit is carried out not only of the scheme itself but also of the surrounding area. Particular attention needs to be given to the tie-ins between the proposed scheme and the existing road network. At the design stages, the site visit may be carried out by one of the Safety Audit Team members.

The Safety Audit must be systematic and objective

Often a number of different plans and schedules need to be reviewed for a single highway scheme. It is imperative that *all* plans and other information (for example, traffic flows, accident details, scheme brief, departures from Standard) are thoroughly examined to check for possible safety problems. Items on different plans must be checked for consistency. At later stages during the Safety Audit process, the Safety Auditor should consult reports written at previous stages of the Safety Audit process. These may not have been written by the same Safety Audit Team.

Clients requiring Safety Audits are advised to produce a Safety Audit brief, listing all of the items provided for audit (Fig. 3.3).

Please note the information that we are submitting for the purposes of the Road Safety Audit.

Design brief	
Design checklist	
Departures from Standard	
Scheme plans (list separately if possible)	
Other scheme details (list separately if possible), e.g., signs, schedules	
Accident printout for existing roads affected by the scheme	
Traffic surveys	
Previous Road Safety Audit reports	
Previous Exception Reports	
Start date for construction	
Any other information (list separately)	

Figure 3.3
Safety Audit brief

Many Safety Auditors use checklists describing some of the potential safety issues, but care should be taken not to be restricted by items on the checklists. The checklist should not be used as a substitute for safety engineering experience. It is sometimes the interaction between design elements that can lead to safety problems and not one element in isolation, for example, a junction on a bend on a downhill section.

Checklists can be used as an *aide-mémoire* after the Safety Auditor has considered the scheme from each road user group's point of view.

The Safety Audit must be looked at from *all* road users' perspectives

In addition to the interaction between design elements, one of the other important checks carried out involves assessing the safety of the scheme from different potential road users' perspectives. During the design stages the Safety Auditor has to imagine what it would be like to walk, cycle and drive the scheme. 'Driving' should include cars, vans, trucks, motorcycles and buses. 'Walking' should be considered from the perspective of the elderly, the child, the wheelchair user and those with sight impairment. For some schemes, the equestrian viewpoint should be considered.

Figure 3.4
Equestrians' needs should be considered along with other vulnerable road users

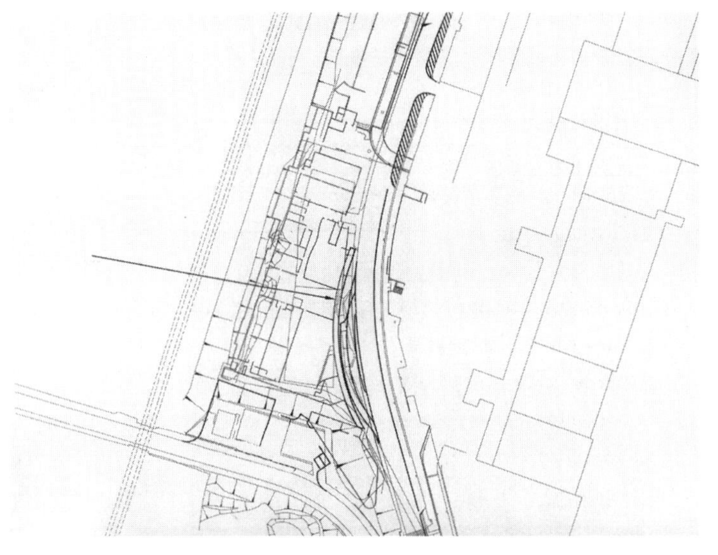

Figure 3.5
Potential conflict between buses leaving the lay-by and other vehicles on the bend

The Safety Auditor tries to develop scenarios when reading the scheme plans, for example, 'what happens if a bus pulls out from this lay-by at the same time as a motorcyclist travels around this bend?' (Fig. 3.5).

Methodology for undertaking Feasibility Stage, Stage 1 and Stage 2 Safety Audits

Safety Audit in the UK is carried out at three key stages on trunk road and motorway schemes[1]. These are Stage 1 (preliminary design), Stage 2 (detailed design), and Stage 3 (pre-opening). There is no requirement to carry out Safety Audit at an earlier or feasibility stage on Highways Agency schemes. Many local authorities do carry out Feasibility (Stage F) Safety Audits, particularly on larger developer led schemes. Stage 1 and Stage 2 audits are often combined on smaller schemes.

The section below describes a working method for carrying out Feasibility, Stage 1 and 2 Road Safety Audits. It is assumed that the Safety Audit Team consists of two members.

(a) The Safety Audit Team looks through plans to understand the scheme concept.
(b) Strong consideration should be given to a meeting between the Safety Audit Team and the design team, particularly on larger or more complex schemes.
(c) One Safety Audit Team member visits the site (taking photographs).
(d) Both Safety Audit Team members systematically and independently examine all plans and other information provided (including photographs taken during the site visit) and write down any comments with reference to plan numbers.
(e) The Safety Audit Team members meet to discuss comments.
(f) The Safety Audit Team decides which comments are related to safety and discusses possible recommendations. Any comments recorded by team members that do not go forward to the final report should be noted, together with a reason stating why that issue is not to be included.
(g) One Safety Audit Team member produces a draft audit report.
(h) The second team member checks the report and edits if necessary.
(i) The Safety Audit Team produces the final report, signs it and sends it to the design team (or client organisation).

A format for recording the Safety Audit Team comments is shown at Fig. 3.6.

SCHEME NAME: Burton Street, Coalville
AUDIT STAGE: Stage 2
DATE: 12 July 2000 **DATE AUDIT REQUIRED BY:** 20 July 2000

AUDITOR'S NAME.....Steve Proctor.....PAGE...1...OF...2...

Plan no.	Road Safety Audit comments	Discussed within audit team	Comment included in report	Reason not included
DBC/HA/601	Downhill approach to traffic signals on 40 mph road – may need anti-skid	yes	yes, para 2.7	
	Diagram number for roundabout warning sign is wrong on plan	yes	no	Not a safety issue as such – point out in letter to client

Figure 3.6

Methodology for undertaking Stage 3 Safety Audits

At Stage 3, it is normal practice for the Safety Audit Team to be accompanied by a police officer and a representative of the organisation that will be responsible for future maintenance. These additional Safety Audit Team members can sign the audit report, but they are not obliged to sign it. A suggested working method for undertaking Stage 3 Safety Audits is as follows:

(a) the Safety Audit Team visits the site during daylight;
(b) the Safety Audit Team walks, drives and, where appropriate, cycles, rides a horse or motorcycle along and across the scheme;
(c) one team member takes notes of all the possible safety points;
(d) the other team member takes photographs of all the possible safety points;
(e) before leaving the site a team meeting is held to ensure that the note-taker has covered all safety points;
(f) the Safety Audit Team visits the site during darkness (this can be carried out by one team member);
(g) one Safety Audit Team member produces a draft audit report and circulates to all present at the site visit;

(h) the report is edited following comments from the other team members;
(i) the Safety Audit Team produces the final report, signs it and sends it to the design team (or client organisation).

There is often pressure to open new road schemes as soon as they are completed. This makes it difficult to carry out the process described above and provide an immediate report to the client. On these occasions it is possible to carry out a 'pre-Stage 3' Audit a couple of weeks before completion. If the recommendations from the pre-Stage 3 Audit are acted upon, the final Stage 3 Audit will be less onerous. It may also be possible to provide the resident engineer with a copy of the handwritten notes taken during the Stage 3 visit. The resident engineer can then start to act upon these notes prior to receiving the formal Stage 3 Audit report.

Figure 3.7
Auditors discuss items on site with the resident engineer during a Stage 3 Road Safety Audit

The Safety Audit report

Having carried out a Safety Audit by looking through scheme plans or examining the completed scheme on site a formal report is written. The following items should be included within the audit report:

- a brief description of the scheme being audited;
- the dates when the Safety Audit was carried out (and the date of the site visit);

- a list of the Safety Audit Team members and any other personnel attending the site visit;
- a series of road safety problems and recommendations for action – it may be useful to include a plan showing the location of the problems;
- a statement signed by the Safety Audit Team members to certify that they have examined the scheme;
- for Feasibility, Stage 1 and 2 Audits, a list of all plans and other information examined.

The main element of the report is the section on problems and recommendations and the following points should be borne in mind when writing this section.

- All problems identified in a Safety Audit report must relate to road safety problems. Non-safety items identified can be itemised in a separate report or letter to the client.
- All safety problems highlighted should be stated as clearly as possible. A clear identification of a problem will help the design organisation to consider not only the recommendation in the report but also to consider alternative ways to overcome the safety problem. An example of a Stage 2 Safety Audit report problem/ recommendation format follows here. In this case a pelican crossing has been proposed on a section of new dual carriageway, shown on the plan in Fig. 3.8.

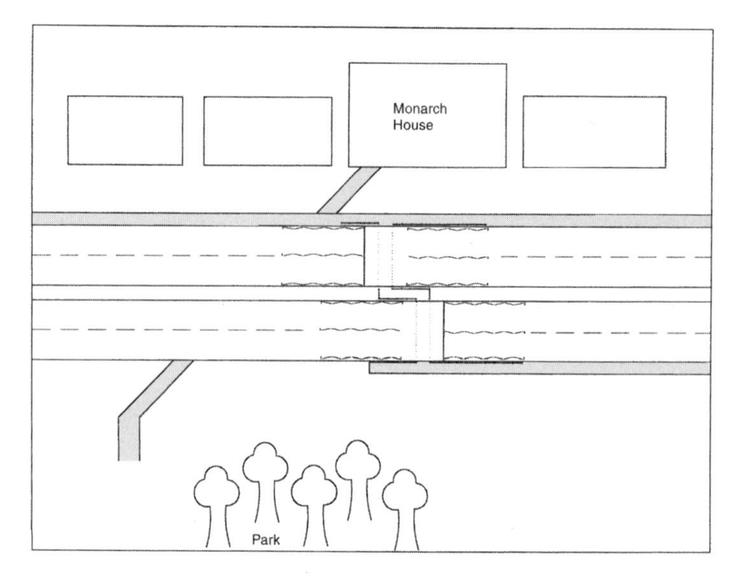

Figure 3.8
Pelican crossing submitted for Safety Audit

> *Problem*
> Pedestrians crossing from the flats to the park may cross the dual carriageway on the exit side of the pelican crossing. Accident studies show that pedestrians are likely to be involved in accidents in these locations.
> *Recommendation*
> Consideration should be given to relocating the pelican crossing on the pedestrian desire line between the flats and the park.

- The Highways Agency requires Safety Auditors to highlight those problems considered to be more severe within the report, but many Auditors are reluctant to do this without using a formal risk assessment procedure. Risk assessment is discussed in more detail in Chapter 8.
- Recommendations should be as practical as possible and be relative to the overall scheme cost. There is little value in putting forward a recommendation that will add more than a small percentage increase to the cost of a scheme. Costs and benefits are discussed in more detail in Chapter 8.
- Safety problems that remain unaddressed throughout the audit process should be repeated at subsequent audit stages. Recommendations may change as appropriate to the stage the design has reached. For example, a Stage 1 Audit on a new high-speed road that crosses an existing footpath has identified a problem of potential accidents as pedestrians cross from one path to another. The recommendation may be to build a footbridge.
  At Stage 2, the Safety Auditors note that the designer has rejected the idea of a footbridge. The problem should be restated, however the recommendation may be that the footpath be re-directed to the nearest over-bridge.

The exception report

The recipient of a Safety Audit report may be the design organisation or the client. Whoever receives the report will have to decide whether or not to act on the recommendations contained within the report. In most cases the client will instruct the designer to make changes in response to the Safety Audit report. Where these are major changes it may be necessary to carry out a re-audit of that part of the scheme.

The most formal way of reacting to the Safety Audit will be through an 'exception report'. The exception report should address all items in the audit report that will not be acted upon. When writing an exception report it should be noted that both the Safety Audit and exception reports could be used in future litigation. The legal implications of Safety Audit are looked at in Chapter 7.

There are two possible types of comments within an exception report. It may be that the client accepts an identified problem, but that the recommendation cannot be implemented for various reasons. In this case, the exception report should describe the alternative measure to be implemented. Discussion between Safety Auditors and designers may help to come up with suitable alternatives, although the present Highways Agency Standard[1] discourages this.

The other scenario is where the recipient of the report does not accept that the identified problem exists. In this case, the exception report should produce some evidence as to why the problem is not valid. It may be that the Audit Team did not have all information available, or that the scheme design has changed since the plans used in the audit were prepared.

Some organisations use a less formal system than an exception report – such as a simple feedback form. The form can be used in response to all audits even when all recommendations are to be adopted.

It is important that a copy of the exception report or feedback form is returned to the Safety Audit Team. Without some form of feedback,

Figure 3.9

SCHEME NAME: Expressway Pelican crossing
AUDIT STAGE: Stage 2
TODAY'S DATE: 20 July 2000

Safety audit paragraph no.	Safety problem accepted	Safety recommendation accepted	Alternative recommendation
2.1	Yes	No	Pedestrian desire line is stronger away from the park so pelican must remain in this location. The path will be re-designed as shown on attached drawing.

Safety Auditors are working in a vacuum and will find it difficult to improve their audits over time. Again, the Highways Agency Standard does not encourage such feedback.

In the example of the pelican crossing referred to above, the feedback shown in Fig. 3.9 was received by the Safety Auditors to their Stage 2 comments. The scheme has been altered as shown in Fig. 3.10.

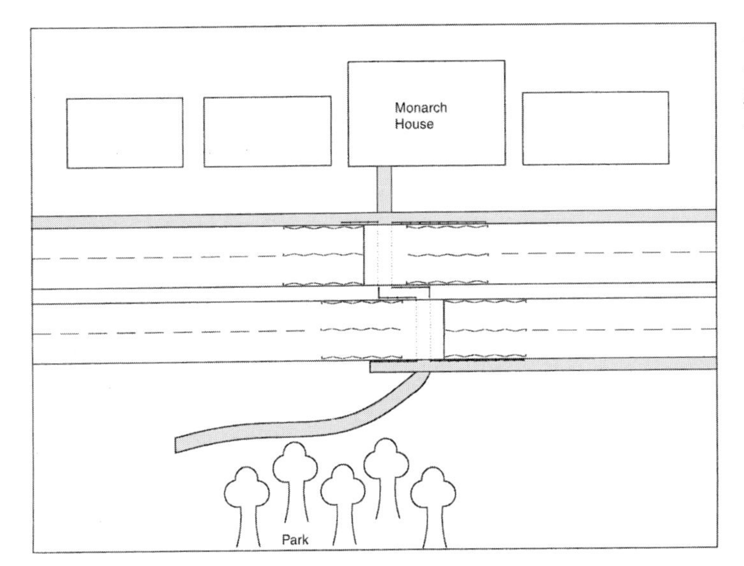

Figure 3.10
Alterations to scheme following Road Safety Audit

In this case the Safety Auditors will have achieved their objective – the safety issue has been addressed, albeit with a different solution.

Arbitration

There will inevitably be some conflict between safety and other issues within the audit process. Some examples are given below:

- large conspicuous road signs are generally a good idea from a safety point of view, while they can have an adverse affect on visual intrusion;
- street lighting generally improves road safety but has implications for light pollution;
- multi-lane approaches to roundabouts can have a poor safety record but will reduce traffic delays.

Figure 3.11
Large roundabouts tend to have entry/circulating accident problems

While the Safety Auditor concentrates on road safety issues, the client or design project engineer will have to weigh up the various consequences of implementing the recommendations within the Safety Audit report. Generally, the project engineer will prepare an exception report or at least a feedback form explaining why recommendations have been rejected. However, occasionally there will be situations where decisions are very difficult and in these cases it may be necessary to introduce a system of arbitration.

The arbitrator should be a senior officer in the organisation commissioning the scheme who has some knowledge of road safety work and who has not been directly involved in the scheme design or the Safety Audit. Arbitration should be a process where both sides present their points of view and the arbitrator makes a decision.

See Fig. 3.12 for a flow chart illustrating the Safety Audit process.

Monitoring

When a scheme has been completed, it is important to monitor its performance in terms of the number and severity of road accidents and casualties. The Safety Audit Standard has a requirement to monitor schemes one year and three years after completion of schemes. The Standard requires that accident locations and common accident types are identified and that accident rates and severity ratios are compared with average rates.

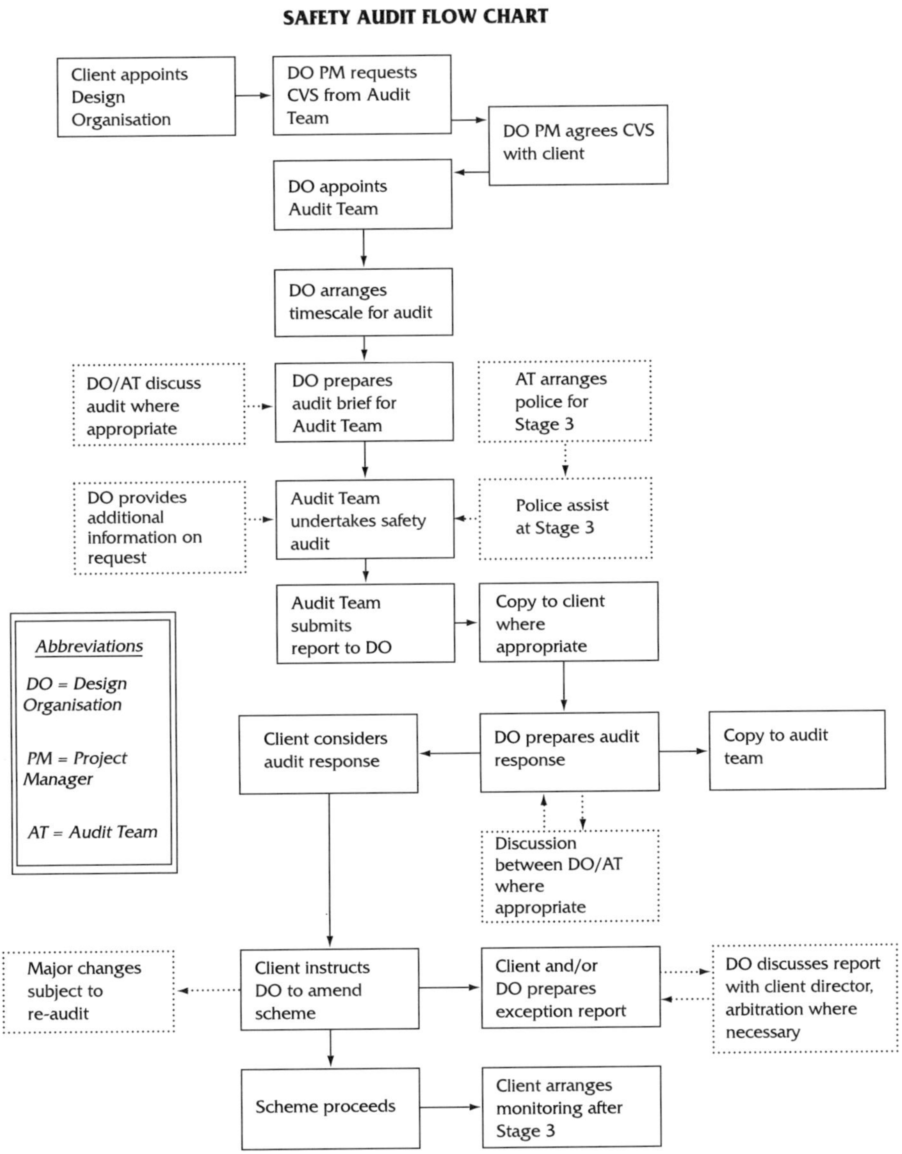

SAFETY AUDIT FLOW CHART

Client appoints Design Organisation

DO PM requests CVS from Audit Team

DO PM agrees CVS with client

DO appoints Audit Team

DO arranges timescale for audit

DO/AT discuss audit where appropriate

DO prepares audit brief for Audit Team

AT arranges police for Stage 3

DO provides additional information on request

Audit Team undertakes safety audit

Police assist at Stage 3

Audit Team submits report to DO

Copy to client where appropriate

Abbreviations

DO = Design Organisation

PM = Project Manager

AT = Audit Team

Client considers audit response

DO prepares audit response

Copy to audit team

Discussion between DO/AT where appropriate

Major changes subject to re-audit

Client instructs DO to amend scheme

Client and/or DO prepares exception report

DO discusses report with client director, arbitration where necessary

Scheme proceeds

Client arranges monitoring after Stage 3

Figure 3.12 *Safety Audit Flowchart*

Although this monitoring is necessary, it is of limited use unless it is related back to the original Safety Audit reports. A suggested working method for monitoring is outlined below:

(a) identify accident locations;

(b) identify accident types;

(c) identify which items highlighted in the Safety Audit report were amended on site;
(d) look at the Safety Audit reports and compare the recorded accidents against the problems highlighted in the report – a comparison of those problems not addressed by the client with accident occurrence will be of particular interest;
(e) prepare a monitoring report for the client (with a copy to the original audit team).

It may also be useful to examine maintenance records as these could highlight where damage-only accidents have occurred.

At the time of writing, the authors are aware of very few genuine Safety Audit monitoring reports being produced in the UK.

Reference

1. HD 19/94 (1994). 'Road Safety Audits', Highways Agency

4. Safety Audit 'control data'

Sources of control data

It is important for Safety Auditors to try to base their comments on sound safety experience, and where possible, to have the means to back up the recommendations from documented sources. The Safety Audit Standard requires the auditor to be able to produce 'background reasoning' for Safety Audits. There are a number of reasons for this.

- To try to avoid road safety 'myths'. Some road safety beliefs are not based on fact. For example, many people believe that the chaos outside schools at entering and leaving time is dangerous (Figs. 4.1 and 4.2). There is very little road accident data to support this 'myth' – in Sandwell in the West Midlands, less than 2% of all injury accidents involve injury to school children on a school journey within 500 metres of a school.
- To try to avoid 'gut feelings' about safety. For example, many engineers believe that certain types of junction control are 'safer' than others, based on personal preference or their design experience rather than accident data.
- To be in a position to substantiate one's case with the client or within arbitration if the audit problem and/or recommendation has been rejected.
- To be in a position to substantiate one's case at a public enquiry on a road scheme that has been subject to Safety Audit.
- To be in a position to substantiate one's case in potential litigation if an accident has taken place on a scheme that has been subject to Safety Audit.
- To avoid wasting time on non-safety issues within the design process.

Figure 4.1
Residential access road approach to infant school

Figure 4.2
The same location ten minutes later

The ideal situation would be one in which we could turn to a published source to answer all of our safety questions. However, this is not always the case. For example, is there data available or do we have to use our own judgement to answer the following questions?

- Is a 'wrong-way' stagger at a dual carriageway pelican crossing unsafe?

Figure 4.3
Pedestrians walk through the central reserve of this pelican crossing with their backs to oncoming traffic

- Is a new mini-roundabout proposed at an existing crossroads junction safe?

Figure 4.4
Four-arm mini roundabouts tend to have a significantly higher accident frequency from those with three-arms

- How much anti-skid should be laid on an approach to a pedestrian crossing on a fast dual carriageway?

Figure 4.5
Drivers need to stop in time on the approach to high speed crossings

Ideally, the background reasoning or 'control data' would contain information from similar sites to predict:

- accident types – this may be possible in some circumstances;
- accident numbers – this is more difficult.

Data is easier to obtain for more substantial elements of scheme design (for example, junction types) than for smaller elements (for example, the '*x*' height of signs). This is partly because the basic source for much accident research is the police STATS19 form, which contains a limited number of codes relating to each accident that has occurred.

Control data should be used to assist Safety Auditors with both the identification of problems, and with the recommendations for improvement. In the first part, the Auditor is trying to determine who is most at risk in the new layout. In the second part, the Auditor tries to suggest an improvement that has been previously demonstrated to mitigate that risk.

Local control data is very important, as accident performance varies from one part of the country to another. It would not be appropriate to use London data as a basis for comparison in rural Cumbria.

The main sources of data are publications and databases. Anyone carrying out Safety Audits should keep up to date with the following.

Publications

- Highways Agency (HA) – Design Standards and Advice Notes contained in the Design Manual for Roads and Bridges (DMRB). However, a recent review of 91 Standards by TMS Consultancy revealed that only 10% include a separate section that provides good control data for the subject covered by the standard. Around 40% of Standards have a medium to high potential for including information of this type. Standards produced since 1995 tend to be better in this respect.
- Department of the Environment, Transport and the Regions (DETR) – Traffic Advisory Leaflets, and other publications.
- Transport Research Laboratory (TRL) reports – the TRL produce around 15–20 reports each year that are useful to the Safety Auditor.
- Commissioned research – for example the AA Foundation for Road Safety Research has produced several detailed safety studies over the past 10–15 years.

- County Surveyors' Society (CSS) reports – several reports bringing together local authority safety experience on a range of issues including street lighting, road surfacing and traffic signals have been produced.
- Local authority accident studies – several large local authorities in England still produce safety reports on the performance of features such as roundabouts and traffic signals.
- Articles in technical journals such as *Traffic Engineering and Control* (TEC), *The Surveyor*, and *Local Transport Today* (LTT).

Figure 4.6
Published reports are an important source of control data

Databases

- TRL manages a database of safety schemes on behalf of CSS – this is known as MOLASSES. Local authorities supply information to TRL and simple 'before and after' results can be obtained.
- Many local authorities maintain their own 'before and after' monitoring systems.
- TMS Consultancy has a database containing around 850 safety schemes from around the UK. The database calculates percentage reductions in accidents and simple economic benefits of schemes. A sample from this database is referred to later in this chapter.

Computer models

- The TRL traffic design models, ARCADY, PICADY, and OSCADY provide accident predictions for specific layouts. However, the accident data upon which some of the models are based are dated and the predictions should be used carefully.

- The TRL has gone a stage further than the junction models with its SAFENET program for modelling accident numbers throughout a town. Based on a series of TRL studies, the program allows the user to build traffic management options and compare accident predictions. The complexity of the program means that it has little direct use for Safety Audit of small schemes.
- Web sites – Safety Auditors can obtain information from a variety of sources on the Internet. Sites representing vulnerable road users may be of particular interest, for example the RNIB, RNID, the BMF and the British Horse Society all have useful sites. The main criterion here, as with all control data, is to check that the information obtained is relevant to the problem in hand.
- In North America, a considerable amount of work has been carried out on accident prediction models that disaggregate accident rates by geometric parameters. This approach can be useful for feasibility or Stage 1 Safety Audits.

Some sources should not be used as reliable control data for Safety Audits. By and large, this type of information relates to 'one-off' accidents that may have occurred several years ago and never been repeated. The Safety Auditor should be trying to reduce the possibility of predictable injury accidents that have a good chance of occurring several times over a five to ten-year period if action is not taken. Sources of data to be wary of include:

- anecdotal information from politicians, engineering professionals, police officers, or members of the public;
- personal 'experience' – this could turn out to be simple prejudice against certain elements of design;
- single fatal accidents that are not representative of a 'pattern' of accidents;
- local newspaper reports that are generally sensationalist in nature and may not be based on what actually happened.

Safety principles from published sources

This section of the book examines a number of sources of information and presents results from published sources. The information ranges from engineering issues such as alignment and junction type, through to road user issues including mobility impairment. Where possible safety information describes typical accident patterns including the type of road users at risk, any 'before and after' results, and the sources of the information. Reference to accidents throughout this chapter should be taken to mean injury accidents unless otherwise stated.

This information should assist Safety Auditors in both the identification of safety problems and in the recommendation of possible improvements. Key findings found to be useful by the authors are italicised.

Some of the information, for example, that relating to alignment or junction type, is fundamental to the early stages of design and is therefore more likely to be of benefit at Stage F or Stage 1. However, there are sometimes situations in which Safety Auditors restate safety problems at consecutive stages of the process, changing the recommendation to suit the stage that the design has reached.

Some of the information may appear to be in conflict, as studies of similar forms of junction control have produced differing results. Some treatments may produce a benefit for one road user at the expense of another. The person who uses this data is advised to supplement the findings with local information, and to take a balanced view for all road users.

Although this information is a comprehensive search through published UK literature sources it can only be a summary of findings, mainly from the past ten to fifteen years. Some of these findings are based on general safety issues such as speed reduction, but many are specific to individual elements of design.

Many new innovations are not fully researched in terms of accident implications, and Safety Auditors have to make judgements related to accident potential. Where appropriate, requirements for further research are identified within this section.

Finally, the information presented here is the authors' interpretation of original research carried out in the main by others. Many of the documents summarised are lengthy publications and readers are advised to consult the sources quoted in order to derive the greatest benefit from the information.

Carriageway type

This section looks at alignment issues and some of the main characteristics of accidents on urban and rural links.

Horizontal and vertical alignment and cross-section

Critical factors are degree of curvature, gradient, and road width. Combinations of critical elements are of particular concern.

IHT Safety Audit Guidelines (1990)

- *Accidents tend to increase with degree of curvature,* with 430 metres being critical in rural areas. Sight distance is a critical factor.

Figure 4.7
Stage 3 Auditors inspect a bend on a new road

- *Accidents tend to be higher at crests and sag curves, especially downgrade sections, mainly due to differential speeds.*
- *Accident rates are higher where substandard horizontal curvature coincides with vertical crests or sags.*
- The rate of increase of accidents related to traffic volume becomes less as the number of lanes increase.
- On rural two-lane roads there are safety advantages in achieving a minimum lane width of 3.65 metres.
- Roads with narrow shoulders and those with shoulders over 3 metres are associated with increased accident rates.
- Central reserves less than 3 metres wide should be avoided for safety reasons.

Figure 4.8
Large vehicles can overhang the central reserve gap into the outside traffic lane

CSS SAGAR Report 1-95 (January 1995). The safety performance of wide single two-lane carriageways

- *Despite problems with overtaking accidents, wide single carriageway roads have a good overall safety record.* Some of the overtaking problems derive from drivers' perception that a modern wide single carriageway is a dual carriageway road.

Figure 4.9
The over-bridge may contribute to the perception that this is a dual carriageway

- Other problems include departures from standard in alignment, lack of visual references in a wide open environment, and clusters of accidents around single lane dualling sections. Measures to counteract these problems include: two-way traffic signs, extensions to central hatching strips, double white lines, and marker posts.

TD 9/93. Highway Link Design

- There is an increase in accidents with gradients over 4%.

TRL Report 334 (1998). The relationship between road layout and accidents on modern rural trunk roads

- *Roads with hard strips are safer than those without by 16–18%.*

Figure 4.10
Pedestrians at risk walking close to the carriageway

- *Wider single carriageway roads are safer than standard width roads by 22%.*
- *Bends, hills and right turn accesses all lead to slightly increased accident risks.*
- Modern dual carriageway roads are safer by one-half compared to traditional dual carriageways.

Figure 4.11
Modern dual carriageways have a good safety record, particularly in rural areas

- Modern single carriageway roads are safer by one-third compared to traditional single carriageways.
- These improvements are due to widespread use of hard strips, safety fencing, wide single carriageways, and fewer and better designed junctions.

- Dual carriageways are safer than single carriageways.

TRL Report 335 (1998). Accidents on modern rural dual-carriageway trunk roads

- Modern single carriageway trunk roads have an accident rate half that of traditional 'A' class roads.

Urban road links

Just over 70% of accidents in Great Britain take place on roads with 30 and 40 mph speed limits subject to street lighting. The following section looks at accidents in urban areas away from junctions where vulnerable road users are over-represented in accidents.

Figure 4.12
Despite a 40 mph speed limit and street lighting, this urban road link appears more rural in nature to a driver, leading to higher speeds

TRL Report 183 (1996). Non-junction accidents on urban single-carriageway roads

- On built-up roads around 60 000 injury accidents occur away from junctions each year. *Pedestrians are involved in 44% of these accidents.* There are more pedestrian accidents where pedestrians cross from the drivers' nearside than from the offside. *Motor cyclists are involved in 17% of accidents, public service vehicles 10% and pedal cycles 8%.*
- *Rear end shunts account for 15% of accidents.* Shunt and lane changing accidents increase on link sections with zebra crossings.
- Accident severity on 40 mph sections is greater than on two-way 30 mph sections.

Figure 4.13
Urban distributor roads are difficult to treat. Here, a cycle lane is partially blocked by a parked vehicle

TecnEcon Report 1996. Pilot red route accident study

- Casualties on 'red routes' in London have reduced by 9%, but on the rat runs in the red route corridor they have reduced by 20–24%, suggesting that improved conditions on the route itself remove unsuitable traffic from the other roads in the corridor.

Research needs

- The safety implications of advisory and mandatory cycle lanes on urban routes, compared to shared or segregated footways.

Figure 4.14
The cycle and pedestrian route is completely blocked by a hedge

Rural road links

Around 30% of accidents in Great Britain take place on roads with a speed limit of more than 40 mph. Less than 4% of accidents take place on motorways where accident rates are one-third those on other rural roads. Somewhat surprisingly, accident severity is also lower on motorways than on other rural roads.

Figure 4.15 *Motorways have a comparatively good accident record*

Accidents in rural areas tend to be of higher severity than accidents in urban areas. Right-turn accidents are a particular problem on single carriageway sections.

TRL Research Report 321 (1991). Accident reductions from trunk road improvements

- Accident remedial schemes on trunk roads can produce a significant saving of around 40% accidents (50% at junctions), with negligible change in accidents downstream of the sites.
- The greatest benefits in terms of accident reductions are from grade-separated junctions (57%), rural road re-alignment (42%), and rural bypasses (32%).

TRL Research Report 365 (August 1992). Injury accidents on rural single-carriageway roads – an analysis of STATS19 data

- *Compared with accidents in built-up areas, those on rural single-carriageway roads are more severe.*
- Just over half of road accident fatalities in Great Britain occur on rural roads.

Figure 4.16
Emergency services attending a high speed rural accident

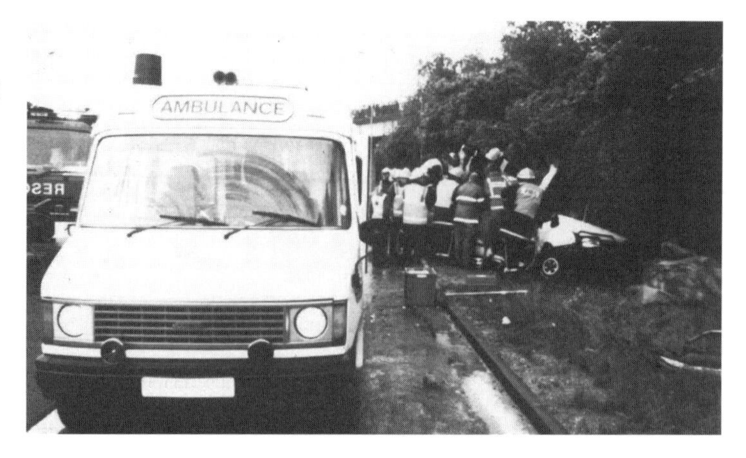

- Most rural injury accidents occur on 'A' roads (53%); on two-lane roads (91%); in 60 mph speed limits (97%); and away from junctions (65%). One in twenty rural accidents occur at private drives.
- The accident severity index[1] for rural single-carriageway roads is 31%. It is higher on wider roads away from junctions, in the dark, and on de-restricted speed limit roads. It is also greater when male drivers, two-wheeled vehicles or heavy goods vehicles are involved.

AA Foundation for Road Safety Research (1994). Accidents on Rural Roads – A study in Cambridgeshire

- *Accidents at strategic single carriageway junctions are dominated by those that involve a right-turn manoeuvre across the line of moving traffic.*
- Junctions are associated with high accident numbers, in particular 'T' or staggered junctions, private drives and slip roads.

Access control and lay-bys

Direct access to high-speed roads can significantly increase accident rates.

IHT Safety Audit Guidelines (1990)

- *Roads with direct frontage access have accident rates twice that of those with limited access.*
- *Where roadside development is extensive, the accident rate may be 20 times as high as that on roads that have access control.*
- *Evidence suggests that there may be a 7% increase in accidents for each additional access point per mile on rural roads.*

[1] Fatal plus serious/total accidents

TD 41/95 (1995). Vehicular access to all purpose trunk roads, Annex 2 – Results on new research on the safety implications of vehicular access

- A significant correlation exists between accidents and traffic flow, link length and farm accesses on single carriageway rural roads. On rural dual carriageways the relationship extends to petrol filling stations.

TA 69/96 (1996). The location and layout of lay-bys

- *One in twenty accidents on rural roads involve parked or stationary vehicles.* The severity index of these accidents is 25%. This highlights the need for adequate access control and provision of lay-bys.
- Accidents at lay-bys represent around 1% of all accidents on rural roads. This includes pedestrians struck while in lay-bys.

Figure 4.17
The combination of signs may lead some drivers to mistakenly assume that the lay-by is an exit slip road – the lay-by is sited in advance of the off-slip

Figure 4.18
Crash evidence in the parking lay-by suggests that this mistake has been made

Junction type

Around 60% of all road accidents in Great Britain occur at junctions. Accidents at major/minor road junctions account for approximately one third of all road accidents in Great Britain. Figure 4.19 illustrates differences between the type of accidents that occur on urban and rural roads. There are more accidents at junctions on urban roads than on rural roads.

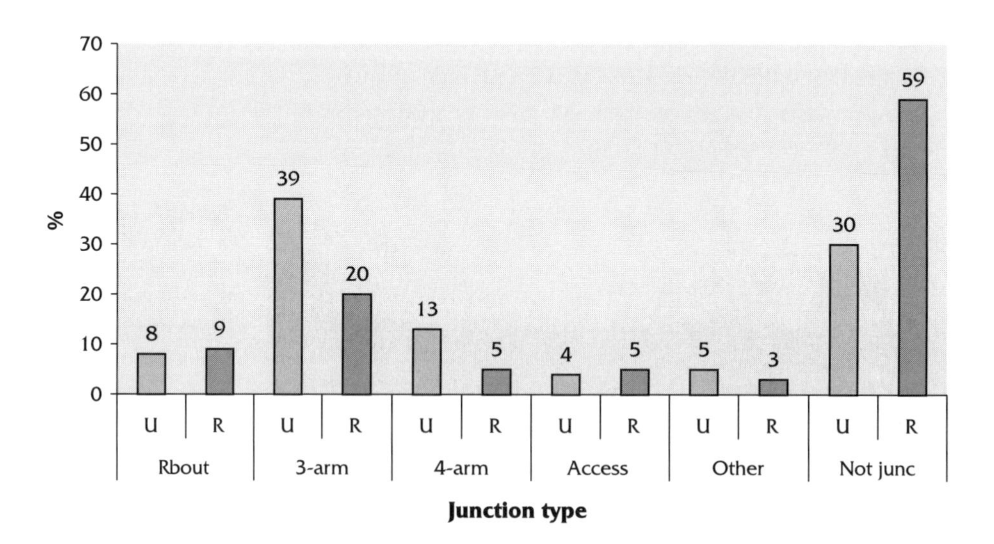

Figure 4.19
Percentage of accidents at urban (U) and rural (R) junctions (source RAGB)

Priority junctions

Priority ('give way' or 'yield') junctions are the most common forms of junction control. Right-turn accidents are the main problem in the UK.

TRL Research Report 65 (May 1986). Accidents at rural T-junctions

- *Motorcycles account for 13% of all accidents even though they only account for 2% of the traffic.*
- *The most frequent junction accident types are those involving a right-turning vehicle from the minor road (27%), a right turn from the major road (22%), and rear-end shunts (20%).*

- *Single vehicle accidents are the most frequent accident type away from junctions (42%), followed by head-on (26%), and rear shunts (20%) respectively.*
- *Downhill approaches to a junction,* on either a major or minor arm, *are associated with higher accident rates.*

TD 40/94. Layout of Compact Grade Separated Junctions

- *An investigation of three-arm priority junctions indicates that 68% of accidents are directly attributable to right-turn manoeuvres,* and 19% to left turn.

TD 42/95. Geometric Design of Major/Minor Priority Junctions

- *Ghost island layouts can reduce right-turn accidents from the major road by up to 70% at three-arm rural T-junctions.*

Figure 4.20
Right turn lanes protect the right turners from potential head-on and rear-end conflict

- *Replacement of a rural crossroads by a stagger can reduce accidents by 60%.*
- *Channelising islands on the minor road approach at rural crossroads can reduce accidents by about 50%.*
- *Conversion of an urban major/minor junction to a roundabout can reduce accidents by about 30%.*

TRL Report 184 (1996). Accidents at three-arm priority junctions on urban single-carriageway roads

- *A 40 mph speed limit on the major road is associated with an increase in accidents.*

- *On average there are more accidents at urban T-junctions with a pedestrian crossing than at those without,* for given vehicle and pedestrian flows.
- The presence of an island on the minor road is associated with increases in all types of accidents, in vehicle-only accidents, in single vehicle accidents and in accidents involving a vehicle turning right into the minor road.
- More traffic lanes increase rear shunt and lane changing accidents on the major arm of the junction, and also increase pedestrian accidents.

TRL Report 185 (1996). Accidents at urban priority crossroads and staggered junctions

- *Longer stagger lengths between minor arms result in fewer accidents of all types,* and fewer vehicle-only and right angle accidents.
- More traffic lanes increase rear shunt and lane changing accidents on a major arm, and also increase pedestrian/vehicle entering accidents.
- A stop line on the minor arm reduces right angle accidents.

Normal roundabout junctions

Roundabouts have a particularly poor safety record for cyclists and motorcyclists. The geometric design of roundabouts is a major contributory factor in the type of accidents that take place.

Figure 4.21
Despite the sign, it is difficult for a driver to spot the roundabout – sited to the left due to constraints on the availability of land

The design of roundabouts in the UK has been based on maximising traffic throughput, which often leads to multi-arm approaches with poor entry path curvature and wide circulatory carriageways, encouraging high speed onto and circulating the roundabout.

TRL LR 1120 (1984). Accidents at four-arm roundabouts

- *Injury accidents involving two-wheeled vehicles constitute 50% of all those reported at four-arm roundabouts on A class roads.*

Figure 4.22 *Motorcyclists are vulnerable at roundabouts*

- *Motorcyclists are involved in 30–40% of all accidents at roundabouts.* The accident involvement rates of two-wheeler riders are about 10 to 15 times those of car occupants.
- *Cyclists are involved in 13–16% of all accidents at roundabouts.*
- *70% of pedal cycle accidents at smaller normal roundabouts are of the entry/circulating type.*

Figure 4.23 *Pedal cyclists are even more at risk!*

- *Pedestrians are involved in 4–6% of accidents at roundabouts.*
- *Accidents at roundabouts of normal design are relatively evenly divided between entering-circulating accidents, approaching accidents and single-vehicle accidents.*

TD 16/93 (September 1993). Geometric Design of Roundabouts

- *The most common problem affecting safety is excessive vehicle speed at entry or within the roundabout.*

Figure 4.24
Wide approach and circulatory carriageways, together with poor entry deflection, lead to drivers taking a 'racing line' through large roundabouts

- *A 57% reduction in accidents can be achieved with the use of yellow bar markings.*

Kent County Council Report (1994). Accidents at roundabouts in Kent

- Motorcyclists are involved in 24–27% of all accidents at roundabouts.
- The accident frequency for roundabouts is 1.0 injury accidents per year.
- Five-arm roundabouts have 1.7 times as many accidents as four-arm roundabouts.

West Midlands Region Road Safety Audit Forum (1997). A Safety Auditor's view of roundabout design

- *Suitable entry path curvature (70 to 100 metres) is the most important safety aspect of roundabout design. A radius of less than 70 metres leads to loss of control accidents on the approach to the roundabout.*

- *Entry angles less than 20 degrees lead to failure to give way and rear end shunt accidents. High entry angles over 60 degrees lead to sideswipe and loss of control accidents.* The ideal entry angle for safety is 30 degrees.

Figure 4.25
Higher entry angles lead to sideswipe and loss of control accidents

- In Warwickshire, the overall accident rate for roundabouts is 1.2 per year. This varied between 0.75 per year for normal roundabouts to 2.03 per year for large roundabouts.
- Narrow lane widths less than 3 metres cause problems for two-wheelers. Wide lanes over 4 metres create poor lane discipline.
- Very large roundabouts lead to high circulating speeds and cause problems for entering traffic.

TA 78/97. Road markings at roundabouts

- *The use of circulatory markings may reduce side-swipe collisions on the circulatory carriageway.* They may also reduce drivers being forced onto the central island, and reduce entry-circulating collisions.

Traffic Advisory Leaflet 9/97. Cyclists at roundabouts, Continental Design Geometry

- *The safety implications of introducing continental roundabout design*[1] *were modelled* on a range of six roundabouts in Oxfordshire and Surrey. Cyclists were involved in accidents at each of the roundabouts. The safety implications were predicted to be:

[1] Continental roundabout design – single lane, straight-on entries, with narrower circulatory carriageways.

- *a reduction in entry/circulating accidents;*
- *a reduction in single vehicle accidents;*
- *an increase in approach arm accidents (shunts).*

Figure 4.26
'Continental' design of a roundabout in Iceland

Transfund New Zealand (2000). The ins and outs of roundabouts

- A useful guide to roundabout safety issues, written in a problem recommendation format. Although the accident data is specific to New Zealand the safety issues of provision for pedestrians, deflection, road markings, signage, visibility, delineation, provision for cyclists, and street furniture location are relevant for most countries.

Research needs

- Actual safety benefits of continental roundabout design in the UK.
- Innovative and safe cycle schemes for cyclists at roundabouts.
- The effectiveness of annular cycle markings at roundabouts. There is some evidence from schemes installed at large roundabouts in the south of England that annular markings may increase pedal cycle accident involvement.

Figure 4.27
Are annular markings the answer to pedal cycle problems at roundabouts?

Mini-roundabout junctions

Mini-roundabouts are often introduced at priority junctions as a traffic safety technique but have a poor record for fail to give way on former main road approaches.

Figure 4.28
Stage 3 Auditors observe drivers ignoring mini-roundabout markings

CSS SAGAR Report 1/4 (May 1987). Small and mini-roundabouts

- *Installation of a small or mini-roundabout at an existing urban priority junction can reduce accidents by between 30 and 40%.*
- *Conversions of normal roundabouts to small or mini-roundabouts can produce increases in accidents of up to 90%.* These results are considered to be a reflection of the effects of inadequate vehicle deflection.

Figure 4.29
Despite the introduction of mini-roundabouts, this road still has the appearance of a main route without a break in priority

- *Four-arm small roundabouts show higher proportions of entry/circulating type accidents than normal roundabouts.*
- *Roundabouts with small central islands and flared entries have accident rates twice those of normal roundabouts.*
- Installation of a small or mini-roundabout at a priority or traffic signal controlled junction can reduce fatal and serious accidents by between 40 and 60%.
- Double mini-roundabouts can reduce accidents at former priority junctions by up to 40%.
- Multiple junctions or double mini-roundabouts installed at sites that were formally normal roundabouts do not reduce accidents.
- Small roundabouts may be more dangerous for cyclists than traffic signals, but small and mini-roundabouts are no worse for pedestrians than other forms of junction control and no different to normal roundabouts for two-wheeled vehicles.

TRL CR 161 (August 1989). Accidents at mini-roundabouts: Frequencies and rates

- *At three-arm mini-roundabouts, the average accident frequency is 0.92 injury accidents per year.* There is a wide regional variation – South West 0.65 accidents per year, North – 0.66 accidents per year, Greater London – 1.70 accidents per year:
  - *pedestrians are involved in 17% accidents;*
  - *pedal cycles are involved in 23% accidents;*
  - *motorcycles are involved in 17% accidents;*
  - *entry/circulating accidents constitute 46% of total accidents.*

Figure 4.30
Drivers on this approach could drive to the right of the mini-roundabout

- *At four-arm mini-roundabouts, the average accident frequency is 1.35 injury accidents per year. Four-arm mini-roundabouts have a 50% greater average accident frequency than three-arm minis.* Again, there is a wide regional variation in the accident frequency – South West 0.78 accidents per year, Greater London 1.70 accidents per year:
  - *pedestrians are involved in 12% accidents;*
  - *pedal cycles are involved in 20% accidents;*
  - *motorcycles involved in 17% accidents;*
  - *entry/circulating accidents account for 66% of the total.*

TRL Report 281 (1998). Accidents at urban mini-roundabouts

- *Accident involvement rates are much higher for pedal cycles and motorcycles.*

Figure 4.31
A cycle by-pass has been installed on the right hand side of this mini-roundabout

- *The average accident severity of accidents at mini-roundabouts is much lower than at priority junctions or at signalised junctions.*
- Pedestrian accidents form a low proportion of the accident total at mini-roundabouts.
- The type of central island does not appear to affect accidents.

West Midlands Region Road Safety Audit Forum (1997). A Safety Auditor's view of roundabout design

- Mini-roundabouts are an accident remedial tool with reduced levels of standards. They can generate accidents and should only be used where there are more than four injury accidents in three years or as part of a traffic calming scheme. *Four-arm mini-roundabouts have a worse accident record than three-arm minis.* Over half of the accidents at mini-roundabouts involve failure to give way.

TRL LR 1120 (1984). Accidents at four-arm roundabouts

- More than two-thirds of accidents at small roundabouts are of the entering-circulating type.
- Accident frequencies can be related to traffic flow and roundabout geometry. Small roundabouts in 30 to 40 mph speed limit zones have both higher accident frequencies and higher accident rates than other roundabout types.

Traffic signal junctions

Traffic signal junctions are more prevalent in urban than in rural situations where right turning accidents, and accidents involving pedestrians are particular safety concerns.

TRL CR65 (1986). Accidents at four-arm single carriageway urban traffic signals

- *The average accident frequency at four-arm urban junctions is 2.65 injury accidents per year per junction,* of which 20% are fatal or serious.
- *The most common accident type is that involving a right turning vehicle (33%).* Of the right turning accidents over 80% involve an 'ahead'

vehicle from the opposite arm (principal right-turn accidents) which contributes some 27% of all accidents.

- *Pedestrian accidents account for 29% of all accidents at traffic signal junctions.* Pedestrian accidents have a high accident severity at 24% and in total account for 35% of all fatal and serious accidents.
- *Motorcycles constitute 3.3% of the traffic, but are involved in over 23% of the accidents.* 45% of these are principal right-turn accidents and in 82% of these it is the motorcycle that is going ahead and colliding with the right turning vehicle.
- *Pedal cycles constitute 2% of the traffic flow and are involved in 10% of the accidents. They have a high involvement in left-turn accidents, particularly of the kind when a cyclist going straight ahead is in a collision with a vehicle turning left across it.*

Figure 4.32 *Advanced stop lines provide safety benefits for cyclists at traffic signals*

- Right-angled accidents have the worst average severity with nearly 30% being fatal or serious.

CSS SAGAR Report 1/6 (August 1989). Automatic traffic signals installation

- *There is no guaranteed safety benefit in the installation, modification or replacement of automatic traffic signals.* It is only when specific problems are identified and treated at high-risk locations that accident reductions are likely.

- *Reductions in right-turn accidents are likely from the provision of facilities to reduce right turns including closely associated secondary signals, port to port turning movements, early cut-offs, right turns held on red, two phase changed to split phase, re-positioning of central loops.*
- *Right turns held on red, early cut-offs, and repositioning of central loops may increase pedestrian accidents.*
- *The introduction of Urban Traffic Control and the provision of closely associated secondary signals are unlikely to have a significant effect on total accident levels.*
- *The provision of pedestrian phases is unlikely to have an effect on accident levels.*
- Increasing the number of signal cycles per hour may increase accident levels.

TRL Report 135 (1996). Accidents at three-arm traffic signals on urban single carriageway roads

- *At three-arm urban junctions the average accident rate is 1.68 injury accidents per year* with a wide variation between regions – 0.92 per year in Wales up to 2.70 per year in Greater London.
- *At three-arm signals, motor cycles are involved in 15% of accidents, pedal cycles in 12% and Public Service Vehicles 10%.*
- *28% of accidents involve rear shunt and lane changing.*
- *The presence of a pedestrian crossing does not necessarily contribute to a reduction in pedestrian and vehicle accidents.* When data for three-arm and four-arm signals are combined, the presence of such a facility actually increases pedestrian accidents.
- Accident severity is 18%, one-third of accidents occur in the dark.
- At three-arm junctions right-turn accidents from the major road have not been reduced by the provision of 'late release' signalling arrangements.
- The provision of a separate lane for left-turners on the minor arm can result in a reduction in the total number of accidents.

Figure 4.33
A new set of traffic signals – but there is no controlled provision for pedestrians within the cycle

Price Waterhouse (1996). Cost benefit analysis of traffic light and speed cameras, Police Research Series, Paper 20

- *Accidents have fallen by an average 18% across 254 traffic signal cameras locations* in ten police force regions in the UK.
- This represents an accident reduction of 0.48 accidents per site per year.

London Research Centre (1997). West London Speed Camera Demonstration Project, London Accident Analysis Unit (see also 'Speed and speed limits' for more details on this study)

- A three-year before and after study has been carried out on 85 kilometres of trunk roads in London. A total of 21 speed camera sites and *12 red light camera sites were included in the study.*
- *Accidents have decreased by 12% compared to controls.* Three-quarters of the decrease is directly attributable to the cameras.
- *Accidents involving failing to stop at traffic signals have reduced by 30%.*
- There has been an increase in shunt accidents, but this does not negate the overall benefits of the scheme.

Traffic Advisory Leaflet 3/97 (1997). The MOVA Signal Control System

- *MOVA* (Microprocessor Optimised Vehicle Actuation) installed across 20 trial sites *showed a small overall reduction in accidents* compared to the before period.

- *Where major, high flow, high speed junctions were examined a 30% reduction in injury accidents was found.*
- Trunk road sites with MOVA showed a reduction in red light running.

Other sources

TA 12/81 (1981). 'Traffic Signals on High Speed Roads'

TA 15/81 (1981). 'Pedestrian Facilities at Traffic Signal installations'

LTN 1/98 (1998). 'The installation of traffic signals and associated equipment'

(SAGAR) (August 1990). 'Special Activity Group on Accident Reduction' Report on the effect of traffic signal cameras.

TD 50/99 (1999). 'The Geometric Layout of Signal-Controlled Junctions and Signalised Roundabouts'

Traffic Advisory Leaflet 7/99: The Scoot Urban Traffic Control System

Research needs

- The safety implications of advanced stop lines for cyclists at traffic signals.
- The effect on safety of bus lanes on the approaches to signals.

Signalised roundabout junctions

Roundabouts are often converted to signal control for capacity reasons. There is some evidence of improvements in safety, particularly for two-wheeled vehicles.

CSS SAGAR Report 1/93 (February 1993). Accidents at signalised roundabouts

- *The total number of accidents at signal controlled roundabouts is similar to that before the signals were installed* (on average 5.85 injury accidents per year at each of the roundabout sites before signalisation and 5.79 after the signals had been installed).
- *The traffic information available indicates a 9% increase in flows through the junctions, so this indicates a slight improvement in the accident rate.*
- Although the total number of accidents is similar in the after period, there appears to be some change in the types of accidents occurring. *There is a decrease in the proportion of two-wheeled vehicles, and an increase in rear-end shunts accidents.*

Figure 4.34
'See-through' to a misleading signal head is a potential problem at signalised roundabouts

Figure 4.35
Here, the drivers can see through to a green signal, when they should be looking at a red

Traffic Engineering and Control (February 1995). Cycle accidents at signalised roundabouts

- *For all categories of signalised roundabouts there is a 36% reduction in the number of cycle accidents at the signalised entries.*

CSS Traffic Management Working Group (1997). A review of signal-controlled roundabouts

- *Overall accident reduction of 11%, after signalisation.*
- *Full-time signal operation shows a reduction of 23%, with accident severity reduced by 40%.*
- *Part-time operation shows an overall increase in accidents* (8% reduction when signals on, 66% increase when signals off).

- *Roundabouts signalised for accident reduction purposes show a 46% reduction in accidents,* those signalised for queue control show a 34% reduction in accidents, and those signalised for capacity purposes show a 16% reduction in accidents.
- The greatest accident reductions after signalisation are on roundabouts with low speed limits and large central island diameters.
- At full-time signalised roundabouts accident severity is reduced by as much as 40%.
- The accident frequency at roundabouts with less than five injury accidents per year is likely to increase if signalised.
- The accident frequency at roundabouts with more than 10 injury accidents per year is likely to decrease if signalised.

Grade separated junctions

TA 48/92 (1992). Layout of Grade Separated Junctions

- *Junctions with a lane drop have a higher accident severity than those without,* indicating the need for more effective signing with gantries.
- *Weaving lengths less than one kilometre approaching junctions have an increased accident rate.*
- *Accident rates on loops are nearly five times higher than on links.* The accident severity on motorways loops is higher than on loops on all-purpose roads.

TD 40/94 (2000). Layout of Compact Grade Separated Junctions

- *The number of accidents at three-arm priority junctions can be reduced by 50% by compact grade separation. This figure is 75% for four-arm priority junctions.* There will be a reduction in accident severity associated with compact grade separation.

Highway features

This section looks at the features that engineers add to the highway once the basic alignment and junction type has been determined.

Street lighting

While street lighting does provide safety benefits in some situations, there is sometimes a problem with the optimum location for a column from a lighting point of view. For example, the best way to light a bend is to put columns on the outside, providing 'silhouette' lighting tails across the carriageway towards an oncoming driver. But columns on the outside of bends are vulnerable to being hit and can cause injury in loss of control accidents.

Figure 4.36
Despite the collision, this lamp is still lit!

CSS SAGAR Report 1/9 (January 1990). Street-lighting installations

- *Modification to street lighting at locations with an above average proportion of dark accidents can result in a significant reduction in those accidents.*
- *There is no significant change in daytime accidents* at locations where street lighting modifications have been implemented.
- New or modified street lighting installations identified for reasons other than high risk are unlikely to show significant changes in either dark or light accidents.
- New street lighting installations at high-risk sites in rural areas show significant reductions in dark accidents but a non-significant reduction in total accidents.

Research needs

- The safety implications of different types of street lighting.
- The safety implications of lighting pedestrian facilities through silhouette lighting compared to floodlighting.

Road surfacing

The skid resistance of the road surface has been shown to be an important safety factor, especially when the surface is wet.

Figure 4.37
A wet road surface on a bend is a particular hazard

HD 28/94 (1994). Skidding Resistance

- *Long-term research has established a link between the risk of a skidding accident and the skid resistance of the road surface.* The risk of skidding accidents with a surface friction coefficient (SFC) below 45 is 20 times worse than with an SFC of more than 60. At an SFC below 35 the risk is 300 times greater.

TRL Report 296 (1991). The relationship between surface texture of roads and accidents

- *Skidding and non-skidding accidents, in both wet and dry conditions, are less if the macro-texture is coarse compared to when it is fine.* This is observed at all levels of underlying skidding resistance.
- *The texture depth below which accident risk begins to increase is about 0.7 mm.*

CSS SAGAR Report 1/5 (1987). Road Surface Treatments

- *Calcined bauxite surface dressing (anti-skid) can substantially reduce wet road accidents and skidding accidents (by 40–60%), but has less effect on dry road accidents.*

Figure 4.38
Anti-skid has been applied on the main road approach to a side road as an accident remedial treatment

- *Low cost surface dressings can reduce accidents particularly in the wet.*
- *Resurfacing and super-elevation can substantially reduce wet and dry road accidents.*
- *New wearing courses can reduce both dry and wet road accidents.*
- *Re-texturing may lead to a reduction in accidents.*
- 50 metres anti-skid should be applied on the approaches to traffic signals and pedestrian crossings on 30 mph roads. In higher speed limit areas up to 100 metres should be considered.

Figure 4.39
The length of anti-skid covers the pelican 'zig-zags', but may not be long enough on this high speed dual carriageway

June 2000. 'Getting a Grip'. Surveyor Magazine

- A three-year before and after study of 18 sites treated in 1992 with anti-skid surfacing, found a 26% reduction in injuries.

- A two-year before and after study of sites treated in 1993 with anti-skid surfacing, found a 42% reduction in injuries.

Impact protection and safety fencing

Around 30% of accidents in Great Britain involve single vehicle non-pedestrian collisions. Just over one-fifth of these accidents involve fatal or serious injuries to vehicle occupants. Protection of occupants through provision of safety fence and other features is an important aspect of highway design.

Road Accidents Great Britain

- Around one quarter of the single vehicle loss of control accidents involve a vehicle leaving the road and going on to strike a variety of objects:
  - trees (16%);
  - lamp posts (12%);
  - safety fence (10%);
  - ditches (10%);
  - traffic signs (7%);
  - 'other' permanent objects (39%).

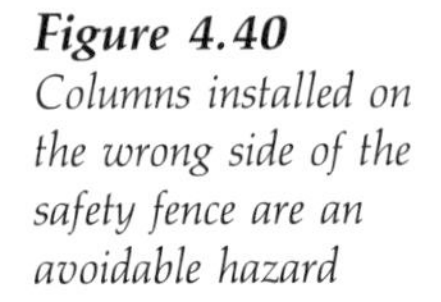

Figure 4.40
Columns installed on the wrong side of the safety fence are an avoidable hazard

TRL RR 75 (1986). Safety Fence Criteria for all purpose dual carriageway roads – a feasibility study

- Accident data for 700 km of dual carriageway with and without central reserve safety fence has been analysed. *The results show a significant cost benefit in installing safety fences in central reserves at traffic flow thresholds below those previously applied.*

- As a result the Highways Agency has completed a ten-year rolling programme of installing central reserve safety fence on trunk road dual carriageways.

Traffic Engineering and Control (June 1995). The relationship between loss of control accidents and impact protection standards: A (TMS) case study

- *It is estimated that injury accidents with wooden fences total around 200 per year.*

Figure 4.41
A fatal accident with a wooden boundary fence on a motorway

- *It is estimated that injury accidents involving vehicles descending embankments total 250 per year.*
- *It is estimated that injury accidents with nearside end terminals total 50 per year.*

Stephen Proctor (1996). 'End treatments to safety fences in Great Britain'. European Road Safety Conference

- *It is estimated that 170 to 300 injury accidents occur each year in GB where vehicles hit exit slip nosing ramped ends.*
- Around 70 to 120 occur each year at motorway ramped ends.
- Birmingham 'crash cushion' installations show a 200% rate of return, with a significant reduction in accident severity.

Figure 4.42
A vehicle has been 'launched' by the ramped end towards the lamp column

Figure 4.43
'Crash cushion' protecting a large sign pole

British Motorcycle Federation briefing note on wire rope safety fence (May 1998)

- Less that 1% of motorcycle collisions are with crash barriers. However, *12% of fatal collisions with barriers involve motorcyclists, due to injuries sustained on exposed posts.*
- Serious injuries suffered by motorcyclists impacting with barriers are caused by the exposed portions of posts on the fencing.

Research needs

- Verification of the findings of the TMS studies quoted above. This information needs to be fed into a new Highways Agency Standard on safety fence.

- Crash tests should be carried out to determine the size of sign that should be protected on a high speed road. Current advice suggests 150 mm posts should be protected, but the installation of safety fence is often avoided by using posts of 148 mm, or using three smaller posts instead of two 150 mm posts.
- Determination of the nature of 'other' permanent objects in RAGB tables.

Pedestrian crossings

While crossings are often regarded as 'safety' features, it is interesting to note that almost as many people are injured crossing roads adjacent to pedestrian facilities as on the crossings themselves.

Some information on pedestrian stages/phases at signals can be found in the section 'Traffic signal junctions'.

CSS SAGAR Report 1/6 (1989). Automatic Traffic Signals
CSS SAGAR Report 1/3 (1987). Pedestrian Crossing Facilities

- *There is no automatic safety benefit in the installation of a new pedestrian facility or in the conversion of one type to another.* Pelicans are not necessarily safer than zebras.

Figure 4.44
Zebra crossings have a comparatively good safety record

- *Pelican and zebra crossings installed at low accident sites can lead to an increase in accidents.*
- *The use of extended green man time can significantly reduce pedestrian accidents at pelicans.*

- *The installation of refuges in the vicinity of pedestrian generators can produce significant reductions in accidents* but care should be taken in the siting of the refuge.
- *Pedestrian guard rail can help to reduce accidents at pedestrian crossings.*

Figure 4.45
Unconventional pedestrian guard rail on a dual carriageway central reserve

- Pelicans are best suited to sites with high vehicle and/or pedestrian flows; accident reductions are most likely at sites with high prior accident rates.
- Pelicans may be safer at night than zebras.
- There may be a difference in pelican/zebra safety with regards to the young and elderly.
- The use of black surrounding 'halo boards' on zebra beacons has not been shown to produce a significant reduction in accidents.

University of Aberdeen, Traffic Engineering and Control (September 1988). Douglas Stewart. Pedestrian guard rails and accidents

- A comparison of visi-rail sites with conventional guard rail showed that *child pedestrian casualties were reduced by 47% at the visi-rail sites, but increased by 90% at conventional sites.* The difference is assumed to be accounted for by the masking of children at conventional guard rails (see Figs 4.46 and 4.47).
- Replacement of conventional guard rail with visi-rail at 12 sites showed all accidents were reduced by 50%.
- Pedestrian accidents reduced by 40%.

Figure 4.46
Children masked by conventional guard rail at a set of traffic signals

Figure 4.47
Good intervisibility provided by high visibility guard rails

Traffic Engineering and Control (February 1990). Pelican crossings and roundabouts in Nottingham

- *Pelican and zebra crossings sited close to the roundabout do not necessarily cause accident problems.*

TRL CR 254 (1991). Accidents at pedestrian crossing facilities

- *Accidents of all types show a reduction of 18% after the introduction of the new facility. Pedestrian accidents show a reduction of 28%.*

Road Accidents Great Britain (1999)

- *10% of pedestrians injured are using a pedestrian crossing, a further 10% are crossing within 50 metres of the facility.*

Figure 4.48
Pedestrians are at risk crossing near to controlled crossings

- 10% of pedestrians injured are walking in the carriageway, 8% are on the footway or verge.

TMS Consultancy research into converting zebra crossings to pelicans (1998)

- Seven schemes were implemented on single carriageway roads, leading to an overall reduction in accidents of 55%, and a reduction in pedestrian accidents of 89%.
- One of the seven schemes showed an increase in accidents, six showed a decrease of 17–79% compared to control levels.

Other sources

TRL Contractor Report 154 (1989). 'Accident rates at zebra and pelican pedestrian crossings in Hertfordshire'

Local Transport Note 1/95 (1995). 'The Assessment of Pedestrian Crossings, DETR'

Local Transport Note 2/95 (1995). 'The Design of Pedestrian Crossings, DETR'

TEC (September 1988). 'Pedestrian guardrails and accidents'

Research needs

- Safety implications of toucan crossings and puffin crossings.

Speed and speed limits

International research suggests a strong link between accident causation, severity, and speed.

Figure 4.49
Misleading speed limit guidance!

DoT (1992) Killing Speed, Saving Lives

- *85% of pedestrians are killed when hit by a car travelling at 40 mph.*
- *45% of pedestrians are killed when hit by a car travelling at 30 mph.*
- *5% of pedestrians are killed when hit by a car travelling at 20 mph.*

Figure 4.50
Damage to vehicle following high speed pedestrian fatality

- *There is an 80% probability of a car occupant being seriously injured in a car that collides with another car or rigid object at 40 mph.*
- *There is a 45% probability of a car occupant being seriously injured in a car that collides with another car or rigid object at 30 mph.*
- *There is a 16% probability of a car occupant being seriously injured in a car that collides with another car or rigid object at 20 mph.*

Figure 4.51 *Intrusion into the passenger cell from a collision with a lamp post*

Traffic Engineering and Control (February 1993). Road safety – Crackdown on speeding – The effect of speed on accident rate

- A reduction in average speed of only 1 mph could cut road deaths by 7%. Speed is a contributory factor in nearly one-third of all road deaths.

TRL PR 58 (1994). Speed, speed limits and accidents

- Inappropriate speed is more of a problem than high speed: 55% of deaths and serious injuries occur in 30 mph limits, 29% in 60 mph limits and 7% in 70 mph limits.

TRL Report 363 (1998). Urban speed management methods

- The report provides information on speed reduction for various measures (see Table 4.1).

Table 4.1. *TRL Report 363: speed reduction for various measures*

Measure	*Effect on mean speed*	*85th percentile*
Traffic calming	−9.3%	−10.4%
Speed cameras	−6.0%	−4.2%
Vehicle activated signs	−4.2%	−4.5%
Flashing signs	−3.8%	no data
Static signs	−2.2%	−3.2%

Figure 4.52
Part-time speed limit operating near a school

Suffolk County Council, unpublished accident data

- Suffolk – Sproughton village – 'Thank you for not speeding' sign shows a 5 mph long-term reduction in 85th percentile speed. Seven injury accidents have occurred in three years before, one injury accident in three years after.
- Suffolk introduced 450 new 30 mph speed limits on main roads through villages. A 4–6 mph reduction in 85th percentile speed was achieved, together with a 20% reduction in accidents.

Price Waterhouse (1996). Cost benefit analysis of traffic light and speed cameras, Police Research Series, Paper 20

- *Accidents have fallen by an average 28% at 475 speed signal cameras installed throughout ten police force regions.*
- *This represents an accident reduction of 1.25 accidents per site per year.*

Figure 4.53
Speed camera installation on a principal road

London Research Centre (1997). West London Speed Camera Demonstration Project, London Accident Analysis Unit

- *Accidents have decreased by 12% compared to controls. Three quarters of the decrease is directly attributable to the cameras.*
- *Accident severity has reduced by 31%,* with fatals reducing by 69%.
- *Accidents involving high or inappropriate speed have reduced by 65%.* Accidents involving failing to stop at traffic signals have reduced by 30%. Single vehicle non-pedestrian accidents have reduced by 30%.
- *There has been a reduction of 41% in pedestrian casualties,* 19% in motor cyclist casualties, and a reduction of 13% in cyclists injured.
- *There has been an increase in shunt accidents,* but this does not negate the overall benefits of the scheme.
- A three-year before and after study has been carried out on 85 kilometres of trunk roads in London. A total of 21 speed camera sites and 12 red light camera sites were included in the study.
- There was no evidence of accident transfer to other roads as a result of the scheme.

Archie Mackie, TRL, Proceedings of the 1998 Speed Management Conference

- *For every 1 mph reduction in mean traffic speed, accidents reduce by 5%.*
- *For every 1 mph reduction in mean traffic speed of 'speeding motorists', accidents reduce by 17%.*

Research needs

- The effectiveness of in-vehicle technology for reducing speeds and following distance.

Traffic calming

Traffic calming, especially road humps, has been proved to be one of the most successful safety measures to be introduced to the UK. Control data presented here refer to the type of accidents that can be treated and the treatments themselves.

Figure 4.54
Very early traffic calming in the Roman city of Pompei

AA Foundation/Birmingham City Council (1990). Accidents to Young Pedestrians: distributions, circumstances, consequences and scope for countermeasures

- *52% of child pedestrian casualties (0 to 15 years) cross the road in which they are injured once a day or more. Only 12% have never crossed the road before.*
- *25% of child pedestrian casualties are injured in the road in which they live, this figure rises to 43% for the 0 to 4 year olds.*
- 46% of drivers involved in child pedestrian accidents have driven through the accident site once a day or more.

Figure 4.55
Location of a fatal child accident – the child lived in this street, and the driver was local

AA Foundation Report (1994) Pedestrian Activity and Accident Risk

- *48% of pedestrian accidents occur on district distributor roads.*

Figure 4.56
Distributor road treated with build-outs

- *29% of pedestrian accidents occur on local distributor roads.*
- *21% of pedestrian accidents occur on residential roads.*

Figure 4.57
New residential estate road treated with road humps – but is conventional traffic calming desirable on a new road layout?

- For distributor roads, 35% pedestrian accidents involve injury to children.
- For residential roads, 54% pedestrian accidents involve injury to children.

TRL Report 312 (1998). Speed cushions

- *Accident reductions are likely to produce savings of 60%.*

Figure 4.58
Three speed cushions of equal dimensions – but the markings make the ramps on the right seem more steep

- The average and 85th percentile speeds at cushions were higher than those measured at 75 mm high flat top and round top humps.
- Narrow (1600 mm) cushions may not provide a sufficient reduction in 20 mph zones.

TRL Report 364 (1998). Traffic Calming on Major Roads

- On the A49 trunk road at Craven Arms, Shropshire, the measures implemented included gateways on each approach, speed cushions, mini-roundabouts, painted 30 mph roundels on patches of red surface, refuges and centre hatching on a red background through the village itself.
- The accident reduction was 45%.

Figure 4.59
Gateway to Craven Arms

Figure 4.60
Speed cushions in Craven Arms

TRL Research Report 263 (February 1990). Urban Safety Project – Overall evaluation of area-wide schemes

- Traffic management measures such as those used within the urban safety project have been effective in reducing accidents by about 13%.
- Accident savings appeared to accrue to all road user groups with somewhat greater benefits to cyclists and motorcyclists.

TMS Consultancy monitoring of local authority safety schemes (see Table 4.2 and also the section on Monitoring later in this chapter)

Table 4.2. *Effectiveness of traffic calming features*

Feature	*% reduction in accidents*
20 mph zones	75
Build-outs	53
Central refuges	43
Chicanes	48
Flat top humps	66
Simple mini-roundabouts	58
Complex mini-roundabouts	33
Horizontal and vertical traffic calming	57
Single pedestrian refuge	20
Narrowings	53
Road humps	68
Round top humps	87
Speed cameras	13
Area traffic calming package	66

Village Speed Control Working Group – Final Report (1994)

- *Gateway signing and traffic calming in villages has resulted in a slight overall reduction in injury accidents (14%)* (see Table 4.3).

Table 4.3. *Speed reduction results*

Type of traffic calming	*85 percentile speed reduction at entrance to village*	*Speed reduction in village*
Minor gateway only	Less than 3 mph	Less than 2 mph
Major gateway only	6–7 mph	2–3 mph
Measures in village alone		3 mph
Gateway and measures in village	Up to 9 mph	Up to 10 mph

Traffic Advisory Leaflet 7/94 (1994). Thumps – Thermoplastic Road Hump

- There were three accidents, two slight and one fatal, in the three-year period prior to installation of Thermoplastic Thumps, and none in the 13-month period afterwards.
- Speed reductions were similar to standard speed humps, but the spacing between thumps was closer, 56 metres compared to 70 metres for road humps.

TRL Report 215 (1996). Review of traffic calming schemes in 20 mph zones

- *The risk of a child being involved in an accident is reduced by about two-thirds where 20 mph zones have been installed.*
- *The average annual accident frequency falls by 60% after the implementation of these schemes.*
- *Child pedestrian and child cyclist accidents fall by 70% and 48% respectively after the schemes were installed,* giving an overall reduction of 67% for all child accidents.
- *Accidents to all cyclists fall by 29%.*
- The average speed at a calming measure is 13.2 mph, and the average speed between calming measures is 17.8 mph. *Overall vehicle speeds fall on average by 9.3 mph.*
- There is a 6.2% reduction in accidents for every 1 mph reduction in speeds.
- Accident migration on the surrounding roads has not been found to be a problem, e.g. Small Heath has one of the largest 20 mph zones in the country – 25 kilometres of treated road. 'Before'

Figure 4.61
Entrance to a 20 mph zone in Small Heath, Birmingham

accidents: 23 injury accidents per year, 'after' accidents: 4.3 injury accidents per year (81% reduction) (source: TMS Consultancy).

TRL Report 186 (1996). Traffic calming – Road hump schemes using 75 mm high humps

- *A 65% average accident reduction is achieved through the installation of 75 mm road humps.*

Figure 4.62
Flat top road hump within 20 mph zone

TRL Report 313 (1998). Traffic calming – an assessment of selected on-road chicane schemes

- *The introduction of chicane schemes results in a 54% reduction in accidents.*

Figure 4.63
Severe chicane scheme on a local distributor road

- The average and 85th percentile speeds at chicanes are reduced by 12 mph compared to the 'before' values. Speeds between chicanes are also reduced.
- The average mean speed at chicanes is about 9 mph higher than the average mean at road humps.

Research needs

- The safety implications of home zones in the UK.

Road markings and signs

There is considerable evidence to suggest that warning signs and road markings can have a significant influence on safety. What is more difficult to determine is the effect of direction signs, and elements such as the '*x*' height of signs.

CSS SAGAR Report 1/8 (June/August 1989). Carriageway definition

- *Chevron signs can reduce accidents at bends by up to 70%.*
- *Hazard marker posts can reduce accidents by between 50% and 70%.*
- *Hatch marking when used as the only measure at urban sites may reduce all accidents by 30%,* but pedestrian accidents may increase.
- Hatch markings together with central refuges can be a beneficial safety treatment, *accidents can be reduced by 50% in rural areas after implementation of hatch markings alone.*

Figure 4.64
The high mounted keep left sign supplements the sign on the bollard within this central refuge

- *Raised rib edge markings on rural motorways can result in significant savings in accidents* involving vehicles leaving the nearside of the carriageway.
- Edge-lining may be beneficial when used as an accident remedial measure at sites with previous loss of control accidents.

Traffic Engineering and Control (December 1993). 3M Traffic Management Group: Short article on the findings of a report (by TMS) investigating the performance of Diamond Grade road-signs

- Eleven sites on high-speed roads treated with Diamond Grade road-signs show a reduction in accidents in the 'after' period. Six of the sites show a statistically significant reduction in accidents compared to control figures. Some of the signs utilised yellow backing boards.

Figure 4.65
Bend treated with chevron mounted on a yellow backing board

TRL PR 49 (1993). Yellow bar markings on motorway slip roads

- *A trial of yellow bar markings applied to 48 motorway slip-roads indicates a reduction in accidents of between 11% and 18%.*

Figure 4.66
Yellow bar markings on motorway off-slip approach to signalised roundabout

TRL PR 118 (1995). M1 chevron trial – accident study

- *Studies reveal a statistically significant reduction in accidents of 56% on and near the chevron markings at two sites on the M1.*
- There is evidence that *the effect persists for 18 km* from the start of the pattern.

- Although the chevrons were specifically designed to influence multi-vehicle collisions, *the largest accident reductions are in the numbers of single-vehicle accidents.*
- Rear-end collisions are reduced at both sites by over 40%.
- The suggestion that chevrons might cause accidents to migrate downstream was investigated, and no evidence for migration has been found.

Figure 4.67
Chevron markings have been used on single carriageway roads as well as motorways

TRL Report (March 1998). Reaching the Limits – The benefit of intelligent signs that target drivers who are at risk or are a hazard to other road users

- Fibre-optic signs are very effective in reducing speeds. They are capable of reducing the number of high-end speeders who contribute to the accident risk, yet without the need for enforcement.

TRL Report 177 (1995). Traffic calming – vehicle activated speed limit reminder signs

- 'Secret' vehicle activated signs show a 2 mph reduction in 85th percentile speed and up to 40% reduction in accidents.

Research needs

- The safety implications of coloured road surface treatments.

Figure 4.68
The long-term safety effects of transverse and central coloured road surfacing is not clear

- The safety implications of distraction caused by advertising hoardings or the use of lamp posts and other street furniture for advertising.

Road works

Major road works are subject to mandatory Safety Audit in Scotland and Wales. Studies show that links with road works have higher accident rates than those without works.

TRL RR 42 (December 1985). Safety performance of traffic management at major roadworks on motorways in 1982

- *The accident rate on motorways with works is 1.5 times higher than on those without works.*
- There appears to be a relatively high risk of accidents associated with operating the secondary traffic on the hard shoulder.
- Accident rates on the cross-overs are similar to those in the contra-flow section and did not constitute a special problem.

TRL RR 223. Study of the safety performance of major motorway roadworks in 1989

- *The report shows 155 personal injury accidents during the periods with major motorway roadworks, compared with 89 in the absence of works.*
- *Personal injury accidents are higher during darkness,* with or without lights than in daylight, both in the works and no-works situation.
- *There was a marked increase in nose-to-tail collisions in the road works sections.*

TRL PR 37 (1993). A review of the accident risk associated with major roadworks on all-purpose dual carriageway roads

- The results of a study of major maintenance schemes suggests that *the effect of these schemes on all-purpose dual carriageways is to increase the personal injury accident rate by 14.5%.*

Traffic Advisory Leaflet 15/99. Cyclists at road works

- Over a five-year period a total of 18 276 (1.6% of total accidents) accidents occurred at road works. Over the same period 950 cycle accidents occurred at road works, representing 0.8% of all cycle accidents.
- The severity of cycle accidents at road works was greater than elsewhere.
- The dominant accident type was a vehicle overtaking a cyclist and striking the rear or offside of the cycle.

Research needs

- Update previous work on accidents at road works.

Vulnerable road user issues

This section examines safety from the point of view of some of those most affected – the vulnerable road users who tend to be more seriously injured when accidents take place.

There is very little information on road traffic accidents affecting some of the most vulnerable road users, for example, people with disabilities and equestrians. Police STATS19 data does not routinely record whether someone is blind, partially sighted, deaf or a wheelchair user. Until 2000, data was not collected on horses involved in road traffic accidents.

The results of an Oscar Faber/DETR study into Junction Improvements for Vulnerable Road Users (the JIVFRU Project) are eagerly awaited. The study has examined around 250 junction treatments designed to offer innovative solutions to pedestrian and/or cycle problems. Although there will be little 'after' accident data, it is anticipated that this study will provide some useful information from before and after conflict studies and speed surveys.

The government's transport policy is to encourage walking, cycling and public transport, particularly for shorter urban trips. This could have a significant short- and medium-term effect on accidents, and must be considered by Safety Auditors when looking at schemes.

Road users with visual impairment

Blind and partially sighted pedestrians are at higher risk when using the highway compared to those who have good vision. Tactile and audible clues are very important to these road users. Some studies have shown that blind pedestrians take up to four times as long to cross the road as sighted pedestrians.

TRL – PR 82 (1995). Accidents involving visually impaired people using public transport or walking

- *37% of the visually impaired have a further handicap. 13% are totally blind.*
- *91% travel alone – this declines with age. One in five would go alone and without directions to a new route, a further 23% would go alone with directions.*
- *29% have been involved in an accident while crossing the road.* Looking at these accidents:
  - *11% are at pelican crossings, 6% are at zebras, 8% are at traffic signals,* 75% are at no designated crossing point.
  - 11% are told it is safe to cross and then struck by a vehicle.
- 37% use a guide dog, 20% use a long cane, 30% use a white stick.

Figure 4.69 *Tactile paving is an essential guide for a blind person. Here the paving stem is too narrow –*

Figure 4.70
– leading to the blind pedestrian stepping over the paving – and missing the crossing

- 79% make daily local trips, 38% make daily town centre trips, and 30% travel outside of their own locality.

Research needs

- As with other road users who have mobility difficulties, there is a real need to obtain and examine road accident data involving sight impairment. The consequences of tactile paving errors would be a good area in which to start this research.

Pedal cyclists

Cyclists are particularly vulnerable at roundabouts, and are over-represented compared to their traffic volume at all types of junctions.

Figure 4.71
Cycle route provided across the bellmouth of an access road

Accidents involving pedal cyclists are particularly under-reported – only one-third of serious accidents and just one-fifth of slight accidents involving cyclists are reported to the police.

Road Accidents Great Britain (1999)

- *Cyclists comprise around 7% of the casualties in Great Britain.*
- *Child cyclists (under 16 years) comprise around 2% of the casualties.*
- *10% of all reported casualties to pedal cyclists occur at roundabouts.*

Figure 4.72 *Cyclists and roundabouts – a dangerous combination?*

TRL Library Services – Current Topics in Transport (1989–1992). Cycling safety

- Riding a bicycle seems to interfere with the speed of mental tasks in older adults. This can be held partly responsible for the high accident involvement of elderly cyclists in complex situations.

AA Foundation for Road Safety Research (1990). The behaviour of teenage cyclists at T-junctions

- *40% of teenage cyclist accidents occur at T-junctions.*
- 10 000 teenage cyclists were injured in cycling accidents in 1988 according to official figures. A hospital study by TRL suggests the actual rate could be three times as high.
- 35% of teenage cyclists appear to disregard the 'rules of the road' for cyclists given in *The Highway Code*, most noticeably in denser traffic and when wishing to proceed across the mouth of the minor

road at T-junctions. Only 2% who fail to comply with *The Highway Code* advice put themselves at real risk.

- Attention is greatest amongst teenage cyclists when turning from a minor road into a major road, but only half could be said to be paying full attention to the road.

TRL PR 15 (1993). Cycling in pedestrian areas

- *It is important not to exclude cyclists from pedestrian areas and force them to use dangerous alternative routes.*
- No real factors justify the exclusion of cyclists from pedestrian areas and this indicates that cycling can be more widely permitted without detriment to pedestrians.

Traffic Engineering and Control (April 1994). Cyclists in danger

- *One-third of incidents cited by pedestrians involved cyclists.*
- Cyclists have a 'near miss' every five miles.

Figure 4.73 *Cyclists riding on footpaths occasionally injure pedestrians – around 300 each year in Great Britain*

Traffic Engineering and Control (February 1995). Cycle accidents at signalised roundabouts

- *For all categories of signalised roundabouts there is a 36% reduction in the number of cycle accidents at the signalised entries.*
- *The results for all roundabouts with full-time signals show an even larger reduction of 66% in accidents to cyclists at the signalised entries.*
- There is no significant change in overall cycle accidents at the roundabouts with part-time signals.

Traffic Engineering and Control (February 1995). A framework for the evaluation of facilities for cyclists – Part 1

- *A segregated cycle path will not automatically result in a reduction in accidents to cyclists.*

Figure 4.74
A plethora of signs to inform cyclists and pedestrians of the status of the footway!

- *Routing a cycle path across, or parallel to tramlines may cause accidents when cycle wheels become trapped in the lines,* or slip on them particularly in wet weather.

Figure 4.75
Cyclist using a road that includes tram tracks

Traffic Advisory Leaflets 8/93, 9/97, 1/97

- *Advanced stop lines (ASLs) can significantly improve safety for cyclists at signal controlled junctions.*
- *Road narrowings to reduce vehicle speed cut cyclist casualties from 1.51 accidents per year to 0.96.*

Figure 4.76
Confusing markings together with fallen leaves may dissuade cyclists from using the feeder to the advanced stop line at the traffic signals

- In Dutch studies the continental design of roundabouts reduced the severity of accidents.

TRL Report 241 (1997). Cyclists at road narrowings

- *Accidents for all vehicles, and accidents involving cyclists either reduce or stay the same after the installation of narrowings.*
- Cyclists are 'squeezed' by motor vehicles overtaking within the narrowing and feel pressured by vehicles following close behind. The solution is to incorporate cycle bypasses and introduce cycle lanes at narrowings.

Traffic Advisory Leaflet 5/97. Cycles and lorries

- *Lorries account for only 2.5% of cycle casualties, but 20% of cycle fatalities.*
- *75% of the fatal accidents involve a lorry turning left across the path of a pedal cyclist, a lorry overtaking a cyclist, or a lorry and cyclist both turning left.*
- Cycle/lorry accidents occur almost exclusively in urban areas.

Research needs

- More research is needed on the road safety effects of an increase in cycling.
- Demonstration projects of cycle solutions at roundabouts should be evaluated and reported.
- Safety evaluation of alternative methods for cycle routes – on-road, segregated/shared footways and off-road routes.

Figure 4.77
Cyclists are signed away from the circulatory carriageway at this roundabout

Bus and tram users

Bus priority schemes are increasing, but as with some of the other vulnerable modes, there is not very much accident-based control data available to the Safety Auditor. Bus casualties constitute less than 1% of the total casualty figures in Great Britain, while around 3% of the total vehicles involved in accidents are buses or coaches.

Figure 4.78
Potential conflict between bus passengers crossing between stops and vehicles could have been alleviated with the provision of a pedestrian crossing

TRL CR 88 (1988). Bus priority by selective detection

- Right-angled accidents (at traffic signals) are found to increase by a factor of 4.5, as cycle time reduces over the range 120 seconds to 32 seconds.
- The benefits of selective detection are relatively modest and can be nullified by even a small increase in accident rates. It is therefore important to consider the safety implications of selective detection, and more weight might be given to maintaining overall cycle times by use of compensation facilities.

TRL CR 180 (1989). Road safety issues for the design of bus priority schemes

- Several sources of road accident problems have been found, particularly in relation to *pedestrian safety in contra-flow bus lanes, and at some pedestrian crossings within both with-flow and contra-flow bus lanes.*
- *Over 20% of accidents appear to involve a vehicle that may have been violating restricted traffic regulations.*
- Other systems such as turning movement concessions for buses or the location of bus stops in bus lanes were shown to have no particular safety problems.

Research needs

- The safety implications of bus lanes and High Occupancy Vehicle (HOV) lanes in particular.
- The safety implications of motorcycle, pedal cycles and taxis using bus lanes.

Figure 4.79
More research is needed on the safety implications of bus lanes

- The safety implications of trams running on-street with other traffic.

Figure 4.80
Conflicts between trams and other users require further investigation

Pedestrians

The government has recently published a document to encourage walking as part of its integrated transport strategy. The number of casualties per kilometre walked has fallen by around 10% compared to the average for 1981–1985. Deaths and serious injuries to pedestrians have fallen by 25%. Part of the reason for the reduction in casualties is a decline in walking. The average distance travelled by pedestrians in Great Britain fell by 19% between 1991 and 1999.

Road Accidents Great Britain (1999)

- *Pedestrians comprise around 13% of the casualties in Great Britain.*
- *Child pedestrians (under 16 years) comprise around 5% of the casualties.*
- *Elderly pedestrians (over 60 years) comprise around 5% of the fatal and seriously injured casualties.*

Figure 4.81
Elderly pedestrians are one of the most vulnerable of road user groups

AA Foundation for Road Safety Research (1990). Accidents to young pedestrians: distributions, circumstances, consequences and scope for countermeasures

- *Two-thirds of all young pedestrian casualties (0–19 years) are travelling to or from home.*
- *One-third are on a trip to or from school* and over one-quarter to or from shops.
- A high proportion of the young pedestrian casualties are masked by parked cars.
- Police blame the pedestrian for the accident on 93% of occasions.
- Pedestrians taking great risks were a factor in some accidents.

AA Foundation for Road Safety Research (1994). Pedestrian activity and accident risk

- *The risk of pedestrian injury in darkness is about four and a half times that in daylight.*
- Department of Trade and Industry figures suggest that up to ten times as many people attend Accident and Emergency Departments of hospitals with injuries sustained in falls on the footways and other transport areas as are injured in road traffic accidents.
- 16% of the walking and 15% of the crossings that take place on district distributors produce nearly 50% of the casualties.

Traffic Engineering and Control (April 1995). Road safety – Why the elderly walk into danger

- *Older people do not understand road safety messages, are confused by some controlled pedestrian crossings and ignore safe islands in the middle of roads.*
- Many older people do not understand the sequence of traffic movements that provides an all green phase for pedestrians to cross.
- Older people are more likely to be at ease if traffic is slowed by calming methods.
- Traffic islands would be of particular help to elderly pedestrians, particularly at bus stops and junctions.

Figure 4.82
Puffin crossings may be confusing to elderly pedestrians who expect to see far side pedestrian signals

AA Foundation for Road Safety Research (1995). Risk and safety on the roads: the older pedestrian

- Although accidents involving pedestrians most often occur on the nearside of the road, *with older people accidents are more likely when the pedestrian is on the far side*. This may indicate faulty judgement of speed and distance in relation to an individual's ability to take evasive action.
- 50% of all pedestrian deaths in Great Britain involve people aged 60 and over.
- Most elderly pedestrian accidents occur in daylight, in fine weather, and in familiar surroundings.
- High rates of pedestrian accidents may be explained in part by drivers paying little attention to pedestrians in their assessment of risk on the road.
- Elderly pedestrians in a state of uncertainty at traffic signals may result in them following others, going elsewhere to cross and trying to anticipate a break in traffic independently of the appropriate phase to cross.

Research needs

- The safety implications of safer routes to school, including 'walking buses'.

Figure 4.83
Many 'safe routes' schemes are implemented in areas with few or low numbers of child casualties

- An evaluation of the safety implications of 'home zones'.
- The safety implications for wheelchair users, for pedestrians with hearing impairment and mental illness.

Figure 4.84
More research is needed on safety implications for wheelchair users

Motorcyclists

Like pedal cyclists, motorcyclist accidents are significantly over-represented at roundabouts and other junctions, compared to their proportion in the traffic flow. There is recent evidence to suggest an accident problem among 'born-again' adult male bikers.

Figure 4.85
Motorcyclists are especially at risk at roundabouts

Road Accidents Great Britain (1999)

- *The total number of motorcyclists injured has increased in each of the last three years, the largest increases being within adult males, who showed a 7% increase in 1999 compared to 1998.*
- *There has been a large reduction in powered two-wheeler casualties in younger age groups compared to the 1981–1985 average* (92% for 17 year olds).
- *There has been an increase in powered two-wheeler casualties in the adult age group compared to the 1981–1985 average* (14% for 25–59 year olds).
- There were 525 powered two-wheeler users killed in 1999 (40% less than the 1981–1985 average).
- There were nearly 25 000 powered two-wheeler users injured in 1999 (58% less than the 1981–1985 average) – 8% of the Great Britain casualty figures.

BMF briefing note on wire rope safety fence (May 1998)

- Less that 1% of motorcycle collisions are with crash barriers. However, *12% of fatal collisions with barriers involve motorcyclists, due to injuries sustained on exposed posts.*

Equestrians

Very little is known about road traffic accidents involving equestrians, who are legally permitted to use all public highways except motorways. Accident data referring to horse riders was collected for the first time in the UK in 1999.

Road Accidents Great Britain (1999)

- 2 horse riders were killed, 33 seriously injured and a further 146 slightly injured in 1999.
- 40% of the casualties were in built-up areas, 60% in rural areas.
- September was the worst month, with 14% of the casualties.

British Horse Society website (October 2000)

- *The British Horse Society believes there are 3000 horse-related road traffic accidents every year.*

Research needs

- More accident-based studies are required to establish the extent and nature of this safety issue.

Monitoring – 'before and after' studies

Traditional road safety engineering work involves identifying high risk locations from accident data, carrying out detailed accident studies at those locations, implementing relevant remedial measures, and then monitoring the effects of those treatments. The results of these monitoring exercises provide useful control data for Safety Auditors – particularly with respect to suggesting recommendations for improvement once a safety problem has been identified.

TMS Consultancy has collected three years 'before and after' data on over 850 safety improvements carried out by local authorities throughout Great Britain. The average cost of the schemes is £13 600, the average reduction in accidents is 34%, and the average cost of saving an accident is £12 600.

Table 4.4 shows an analysis of schemes costing less than £100 000 by scheme category.

Statistical analysis of 'before and after' data

There are a number of reasons why a reduction in accidents might occur at a location, following the introduction of a road safety scheme:

- the scheme has been successful and reduced risk to road users;
- the reduction is due to a change in traffic levels coincident with (or caused by) the scheme;

Table 4.4 *Analysis of schemes costing less than £100 000*

Group description	*Average cost (£)*	*Reduction in accidents (%)*	*Cost of saving 1 accident per year*
Road improvements 38 schemes	15 882	62	11 955
Visibility improvements 12 schemes	7 890	27	17 261
Right turn lanes 12 schemes	11 849	48	6 921
Junction improvements 34 schemes	18 513	44	16 617
Mini-roundabouts 18 schemes	14 769	49	20 862
New traffic signals 15 schemes	40 717	67	18 248
Traffic signal improvements 16 schemes	17 095	22	18 054
Lighting schemes 14 schemes	9 709	21	13 629
Anti-skid surfacing 34 schemes	8 620	57	7 928
Resurfacing 27 schemes	13 810	46	15 920
Pelican crossings 47 schemes	16 922	24	64
Zebra crossings 15 schemes	12 511	39	154
Conversion from Zebra to Pelican crossings 10 schemes	12 470	46	158
Pedestrian crossing improvements 35 schemes	11 057	41	11 632
Pedestrian refuges 65 schemes	10 387	37	10 774
Pedestrian guard rail schemes 28 schemes	6 230	30	6 009
Speed cameras 28 schemes	18 236	13	10 737
Area-wide traffic calming schemes 14 schemes	46 093	57	12 901
Horizontal traffic calming schemes 16 schemes	22 606	46	22 364

Table 4.4 *Continued*

Group description	*Average cost (£)*	*Reduction in accidents (%)*	*Cost of saving 1 accident per year*
Vertical traffic calming schemes 58 schemes	23 333	65	14 111
Chevron signs 14 schemes	2 505	43	3 062
Warning signs 36 schemes	553	46	799
Other road signs 30 schemes	1 691	50	2 134
Road signs and markings 63 schemes	2 537	41	3 403
Road markings 43 schemes	2 020	34	2 915
Packages of measures 97 schemes	22 099	42	16 304

- the reduction is due to the transfer of risk elsewhere coincident with (or caused by) the scheme, sometimes referred to as 'accident migration';
- the reduction is due to changes in local or national trends in accident levels that are reflected throughout the locality, not just at the treated site;
- the reduction is due to random fluctuation in accident numbers, sometimes referred to as 'regression to the mean'.

In order to determine whether a real reduction in risk has occurred, the monitoring process can attempt to make allowance for the possible reasons for the apparent accident reduction. Traffic levels can be observed, and accidents at adjacent sites can be checked to determine the extent of accident transfer. Comparisons of site data with control data using Chi Squared testing can make allowances for local or national trends.

The phenomenon of regression to mean needs to be considered as part of this analysis. The theory of regression to mean implies that because locations are chosen for investigation when the accident total is high, accident levels will come down regardless of treatment, due to random fluctuation.

However, it is likely that many locations with a consistent pattern of accidents over a long period of time are predisposed to accident occurrence through some highway factor interacting with human behaviour. In addition, there may have been changes to the highway in previous years that may have influenced accident levels but have gone unrecorded, for example, routine maintenance of carriageway markings.

Furthermore, safety schemes are often implemented some time after the identification of problems, due to budgetary or other constraints. This leads to the possibility that accident locations are chosen when numbers are high, after which the numbers come down through regression to mean. A delay in implementation occurs, during which time accident numbers start to increase again through regression to mean. The treatment is then imposed against a background of increasing rather than decreasing accident frequencies.

This is clearly a complex area, and it is perhaps not surprising that very little routine analysis to take account of regression to mean has been carried out by UK local authorities when monitoring the results of their safety schemes. One method suggested involves the use of 'matched pair' controls, where for each treated location, an identical non-treated location is selected, and monitored. There are two problems with this approach. First, it is almost impossible to find a 'matched pair', in terms of identical layout, traffic flows, traffic mix, accident pattern and frequency. Secondly, even if it were possible to do this, there could be serious legal implications of leaving locations with identified accident problems untreated.

As far as this book is concerned, the potential for regression to mean is acknowledged, along with the acceptance of the possibility that some of the claimed accident reduction values will overstate the real change that has occurred.

References

Design Manual for Roads and Bridges

TA 12/81 (1981). 'Traffic Signals on High Speed Roads'

TA 15/81 (1981). 'Pedestrian Facilities at Traffic Signal installations'

TA 49/86 (1986). 'Appraisals of New and Replacement Lighting on Trunk Roads and Trunk Road Motorway'

TA 48/92 (1992). 'Layout of Grade Separated Junctions'

TD 9/93 (1993). 'Highway Link Design'

TD 16/93 (1993). 'Geometric Design of Roundabouts'

HD 28/94 (1994). 'Skidding Resistance'

TD 40/94 (1994). 'Layout of Compact Grade Separated Junctions'

TD 41/95 (1995). 'Annex 2 – Results on new research on the safety implications of vehicular access'

TD 42/95 (1995). 'Geometric Design of Major/Minor Priority Junctions'

TA 69/96 (1996). 'The location and layout of lay-bys'

TA 78/97 (1997). 'Design of Road markings at Roundabouts'

TD 50/99 (1999). 'The Geometric Layout of Signal Controlled Junctions and Signalised Roundabouts'

Transport Research Laboratory Reports

TRL (1985). Research Report 42, 'Safety performance of traffic management at major roadworks on motorways in 1982'

TRL (1984). Laboratory Report 1120, 'Accidents at 4-arm roundabouts'

TRL (1986). Research Report 65, 'Accidents at rural T-junctions'

TRL (1986). Report CR65, 'Accidents at four-arm single carriageway urban traffic signals'

TRL (1986). Research Report 75, 'Safety Fence Criteria for all-purpose dual carriageway roads – a feasibility study'

TRL (1988). Contractor Report 88, 'Bus priority by selective detection'

TRL (1989). Contractor Report 154, 'Accident rates at zebra and pelican pedestrian crossings in Hertfordshire'

TRL (1989). Contractor Report 161, 'Accidents at mini-roundabouts: Frequencies and rates'

TRL (1989). 'Study of the safety performance of major motorway roadworks in 1989 RR 223'

TRL (1996). Report 183, 'Non-junction accidents on urban single-carriageway roads'

TRL (1989). Contractor Report 180, 'Road safety issues for the design of bus priority schemes'

TRL (1998). Report 281, 'Accidents at urban mini-roundabouts'

TRL (1997). Report 285, 'Continental roundabouts and cyclists'

TRL (1991). Report 296, 'The relationship between surface texture of roads and accidents'

TRL (1992). Library Services – 'Current Topics in Transport, Cycling safety' (1989–1992)

TRL (1990). Research Report 263, 'Urban Safety Project Overall evaluation of area-wide schemes'

TRL (1991). Contractor Report 254, 'Accidents at pedestrian crossing facilities'

TRL (1991). Research Report 1991, 'Accident reductions from trunk road improvements'

TRL (1993). Project Report 15, 'Cycling in pedestrian areas'

TRL (1993). Project Report 37, 'A review of the accident risk associated with major roadworks on all-purpose dual carriageway roads'

TRL (1992). Research Report 365, 'Injury accidents on rural single-carriageway roads – an analysis of STATS19 data'

TRL (1993). Project Report 49, 'Yellow bar markings on motorway slip-roads'
TRL (1994). Project Report 58, 'Speed, speed limits and accidents'
TRL (1995). Report 177, 'Traffic calming – vehicle activated speed limit reminder signs'
TRL (1995). Project Report 118, 'M1 chevron trial – accident study'
TRL (1995). Project Report 82, 'Accidents involving visually impaired people using public transport or walking'
TRL (1996). Report 215, 'Review of traffic calming schemes in 20 mph zones'
TRL (1996). Report 135, 'Accidents at three-arm traffic signals on urban single carriageway roads'
TRL (1996). Report 184, 'Accidents at three-arm priority junctions on urban single-carriageway roads'
TRL (1996). Report 185, 'Accidents at urban priority crossroads and staggered junctions'
TRL (1998). Report 335, 'Accidents on modern rural dual-carriageway trunk roads'
TRL (1998). Report 334, 'The relationship between road layout and accidents on modern rural trunk roads'
TRL (1998). Report, 'Reaching the Limits – The benefit of intelligent signs that target drivers who are at risk or are a hazard to other road users'
TRL (1998). 'Proceedings of the Speed Management 98 Conference, Archie Mackie'
TRL (1985). Report 85, 'Speed Reduction in 24 Villages'
TRL (1995). Report 177, 'Traffic calming – vehicle activated speed limit reminder signs'
TRL (1995). Report 182, 'Traffic calming – four schemes on distributor roads'
TRL (1996). Report 186, 'Traffic calming – road hump schemes using 75 mm high humps'
TRL (1997). Report 241, 'Cyclists at road narrowings'
TRL (1998). Report 312, 'Speed cushions schemes'
TRL (1998). Report 313, 'Traffic calming – an assessment of selected on-road chicanes schemes'
TRL (1998). Report 363, 'Urban speed management methods'
TRL (1998). Report 364, ' Traffic Calming on Major Roads'

Special Activity Group on Accident Reduction (SAGAR) Reports
SAGAR (January 1995). Report 1-95, 'The safety performance of wide single two-lane carriageways'
SAGAR (May 1987). Report 1/4, 'Small and mini-roundabouts'
SAGAR (1987). Report 1/5, 'Road Surface Treatments'
SAGAR (August 1989). Report 1/6, 'Automatic traffic signals installation'
SAGAR (1987). Report 1/3 'Pedestrian Crossing Facilities
SAGAR (June/August 1989). Report 1/8, 'Carriageway definition'
SAGAR (January 1990). Report 1/9, 'Street-lighting installations'

SAGAR (August 1990). Report 1/12. 'The effect of traffic signal cameras'

SAGAR (February 1993). Report 1-93, 'Accidents at signalised roundabouts'

SAGAR (August 1993). Report 7-93, 'The use of cameras for the enforcement of speed limits: enhancing their effectiveness'

A review of signal-controlled roundabouts 1997, CSS Traffic Management Working Group

Automobile Association Foundation for Road Safety Research Reports

AA (1990). Foundation/Birmingham City Council, 'Accidents to Young Pedestrians'

AA (1994). Foundation Report, 'Pedestrian Activity and Accident Risk'

AA (1994). Foundation Report, 'Accidents on Rural Roads'

AA (1990). Foundation for Road Safety Research, 'The behaviour of teenage cyclists at T-junctions'

AA (1990). Foundation for Road Safety Research, 'Accidents to young pedestrians: distributions, circumstances, consequences and scope for countermeasures'

AA (1994). Foundation for Road Safety Research, 'Accidents on Rural Roads – A study in Cambridgeshire'

AA (1994). Foundation for Road Safety Research, 'Pedestrian activity and accident risk'

AA (1995). Foundation for Road Safety Research, 'Risk and safety on the roads: the older pedestrian'

Traffic Advisory Leaflets (TAL) and Local Transport Notes (LTN)

TAL (9/97). 'Cyclists at roundabouts, Continental Design Geometry'

TAL (2/90). 'Speed Control Humps'

TAL (3/90). 'Urban Safety Management'

TAL (7/91). '20-MPH Speed limit zones'

TAL (2/92). 'The Carfax, Horsham 20 MPH zone'

TAL (7/93). 'Traffic Calming Regulations'

TAL (12/93). 'Overrun Areas'

TAL (13/93). 'Gateways'

TAL (1/94). 'VISP – A Summary'

TAL (2/94). 'Entry Treatments'

TAL (7/96). 'Highways Regulations 1996'

TAL (1/97). 'Cyclists at Road Narrowings'

TAL (2/97). 'Traffic Calming on Major Roads – A49, Craven Arms and Shropshire'

TAL (6/97). 'Traffic Calming on Major Roads – A47, Thorney and Cambridgeshire'

TAL (12/97). 'Chicane Schemes'

TAL (1/98). 'Speed Cushion Schemes'

TAL (4/99). 'Traffic Calming Bibliography'

TAL (9/99). '20 MPH Speed Limits and Zones'

TAL (14/99). 'Traffic Calming on Major Roads: A Traffic Calming Scheme at Costessey, Norfolk'
TAL (7/94). 'Thumps – Thermoplastic Road Humps'
TAL (3/90). 'Urban Safety Managment'
TAL (4/90). 'Tactile Marking for Segregated Shared Use By Cyclists and Pedestrians'
DETR (1/95). LTN, 'The Assessment of Pedestrian Crossings'
DETR (2/95). LTN, 'The Design of Pedestrian Crossings'
DETR (1/98). LTN, 'The installation of traffic signals and associated equipment'
TAL (3/97). 'The "Mova" Signal Control System'
TAL (7/99). 'The Scoot Urban Traffic Control System'
TAL (15/99). 'Cyclists at road works'
TAL (5/97). 'Cycles and lorries'
TAL (10/00). 'Road humps. Discomfort, noise and ground-borne vibration'
TAL (11/00). 'Village traffic calming – reducing accidents'
TAL (1/00). 'Traffic calming in villages on major roads'

Magazine articles – Traffic Engineering and Control (TEC)
TEC (September 1988). 'Pedestrian guardrails and accidents'
TEC (February 1990). 'Article Notts'
TEC (February 1993). Road safety – Crackdown on speeding – 'The effect of speed on accident rate'
TEC (November 1993). Road safety – 'Speed cameras save lives'
TEC (December 1993). 3M Traffic Management Group. 'Short article on the findings of a report investigating the performance of Diamond Grade road-signs'
TEC (February 1993). 'Advanced stop-lines for cyclists'
TEC (April 1994). 'Speed cameras' dramatic effect'
TEC (April 1994). 'Cyclists in danger'
TEC (April 1995). Road safety – 'Why the elderly walk into danger'
TEC (February 1995). 'A framework for the evaluation of facilities for cyclists – Part 1'
TEC (February 1995). 'Cycle accidents at signalised roundabouts'
TEC (June 1995). 'The relationship between loss-of-control accidents and impact protection standards' A case study
Surveyor Magazine (June 2000). 'Getting a Grip'
TEC (September 1988). Douglas Stewart. 'Pedestrian guard rails and accidents'

Other references
West Midlands Region Road Safety Audit Forum (1997). 'A Safety Auditors view of roundabout design'
Kent County Council (1994). 'Accidents at roundabouts in Kent'
Accidents at mini-roundabouts in Kent (1980 to 1987 inclusive). Kent County Council

'BMF briefing note on wire rope safety fence' (1998) British Motorcycle Federation website

DETR (2000). 'Encouraging Walking'

S Proctor (1996). 'End Treatments to safety fences in GB'

IHT (1990). 'Safety Audit Guidelines'

DoT (1992). 'Killing Speed, Saving Lives'

London Research Centre: 'West London Speed Camera Demonstration Project' London Accident Analysis Unit

National Cycling Forum (June 1998) 'Issues for Traffic Engineers and Transport Planners'

DETR (1996). 'National Cycling Strategy'

DETR (1998). 'Places, Streets and Movement'

Price Waterhouse (1996). Cost benefit analysis of traffic light and speed cameras, Police Research Series, Paper 20

Suffolk County Council, unpublished accident data

TecnEcon Report (1996). Pilot red route accident study

TMS Consultancy monitoring of local authority safety schemes

Village Speed Control Working Group – Final Report 1994

Transfund New Zealand (2000). 'The ins and outs of roundabouts'

TMS Consultancy research into zebra/pelican conversions (1998)

5. Problems identified within Safety Audit reports

Chapter 4 of this book examined published sources of safety information useful to the Safety Auditor. This chapter concentrates on the type of problems identified by experienced Safety Auditors, with particular reference to studies carried out by TMS Consultancy.

TMS Safety Audit experience

Between 1991 and 2000 TMS Consultancy carried out over 1500 Safety Audits on a wide range of schemes, considered to be representative of Safety Audits carried out in Great Britain. The Safety Audits were carried out by eight auditors working in a variety of two-person teams. The previous experience of the TMS auditors was predominantly road safety, traffic and highway engineering. Figs 5.1 to 5.4 provide a general analysis of this work.

The information contained within the TMS Safety Audits forms the basis for most of the research described in this chapter.

Study of problems referred to in Safety Audits

General description of audit comments

In 1999, TMS Consultancy carried out a study of a sub-set of their Safety Audits, together with a set of Safety Audits carried out by other

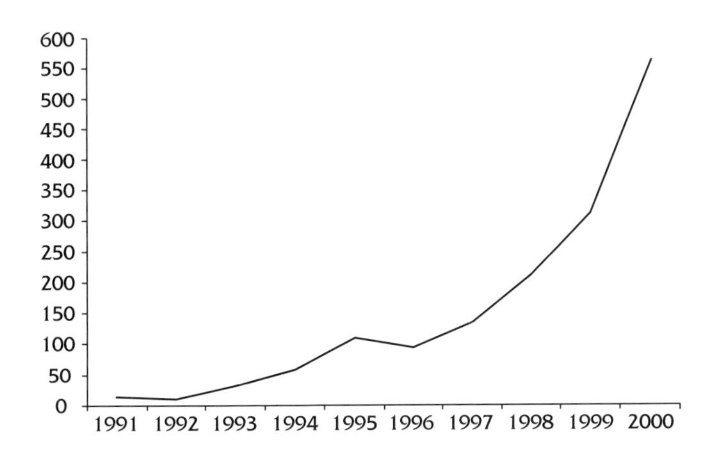

Figure 5.1
Number of TMS audits over time

Figure 5.2
Audits by Stage

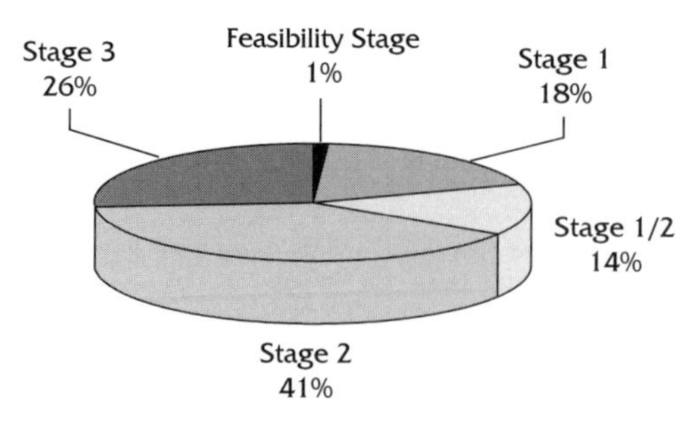

Figure 5.3
Audits by Client

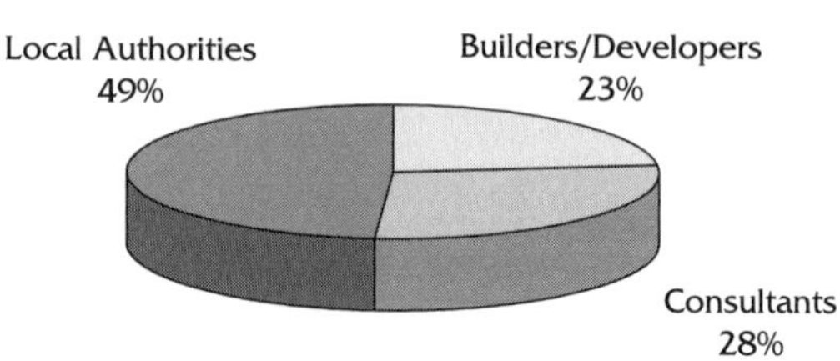

consultants made available by the Highways Agency. A total of 113 Safety Audits carried out between January 1997 and May 1999 were reviewed. The database of Safety Audits included information from 30 Safety Audit reports from the Highways Agency and 83 reports from TMS. The 113 reports provided a total of 1571 separate Safety Audit comments (problem/recommendation statements), an average of 14 comments per report.

The distribution of the comments by audit stage is broadly in line with overall TMS experience of Safety Audits and is described in Fig. 5.5.

Over one-third of the comments referred to vulnerable road users, as described in Fig. 5.6.

A total of 233 comments referred to pedestrian issues. Of these comments, 12% related specifically to pedestrians with sight impairment, and a further 11% to those with mobility impairment.

Although a very small number of comments referred specifically to motorcyclists, many of the issues identified for vehicle users would also apply to this group.

Figure 5.4
Audits by road type

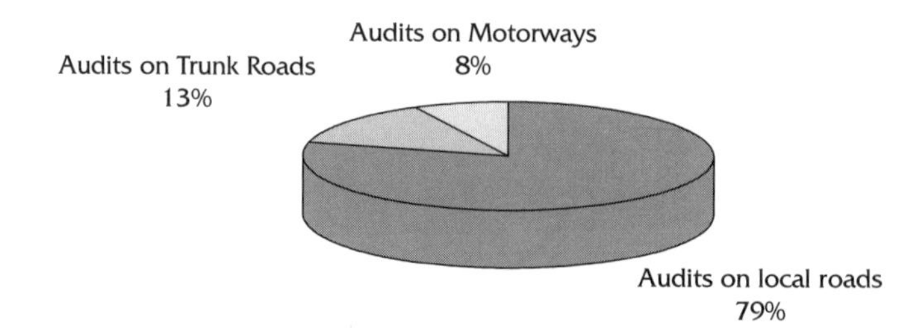

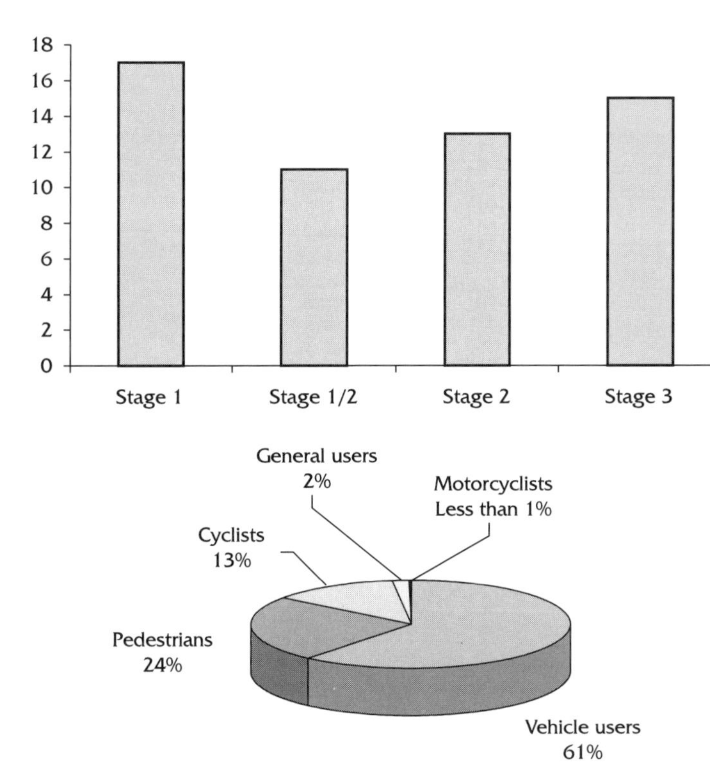

Figure 5.5
Average number of comments per report by stage of audit

Figure 5.6
Number of comments for each road user

Just under half of the comments related to schemes on links of road; these links include controlled pedestrian crossing facilities. Roundabouts and mini-roundabouts together made up one-fifth of the comments, traffic signals one-tenth. The spread of schemes is shown in Fig. 5.7.

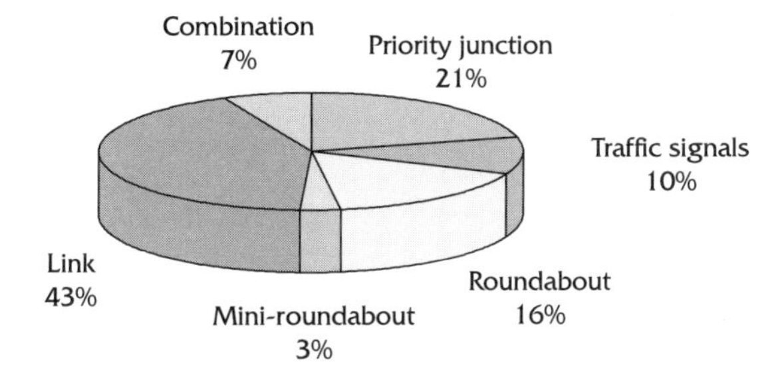

Figure 5.7
Proportion of different types of schemes in TMS study

Specific problems revealed within the audit reports

Table 5.1 and Fig. 5.8 show the individual issues leading to potential safety problems that were identified in the study on ten or more occasions within Safety Audit reports.

The vast majority of the comments (97%) related to issues that the Safety Auditors believed could lead to road safety problems. A minority

***Table 5.1** The most common issues identified*

Description of problem	*count*	*% of total*
1. Inadequate road signs	207	13.6
2. Inadequate road markings or road studs	156	10.3
3. Visibility to signs or traffic signals restricted	80	5.3
4. Inadequate tactile paving	71	4.7
5. Problems with traffic lanes – number or width	64	4.2
6. Inadequate lighting or reflectivity of signs	62	4.1
7. Unsafe crossing point for vulnerable road users	53	3.5
8. Problem for future maintenance	50	3.3
9. Inadequate provision for vulnerable road users	47	3.1
10. Inadequate safety fence	47	3.1
11. Dropped crossing absent or not flush	38	2.5
12. Poor design or absence of splitter island or refuge	35	2.3
13. Risk of actual speed being greater than design speed	34	2.2
14. Stopping problem – absence or insufficient length of anti-skid	34	2.2
15. Footway obstructed by poorly-sited street furniture	33	2.2
16. Visibility to pedestrians restricted through poor guard rail or planting	31	2.0
17. Lack of continuity of a facility	30	2.0
18. Obsolete road signs or markings	30	2.0
19. Specific feature inconspicuous	30	2.0
20. Forward visibility restricted	29	1.9
21. Set back from kerb edge too small	29	1.9
22. Drainage problems and location of gullies	28	1.8
23. Horizontal alignment too severe	27	1.8
24. 'See through' problem	26	1.7
25. Pedestrian guard rail absent or insufficient	24	1.6
26. Absence of pedestrian crossing points	21	1.4
27. Poorly-sited street furniture	20	1.3
28. Junction control	19	1.3
29. Conflict between different road users	19	1.3
30. Poorly designed traffic calming feature	18	1.2
31. Poor design of bus stops	16	1.1
32. Previous accident history unaccounted for	14	0.9
33. Loss of control hazard	13	0.9
34. Parking problem	11	0.7
35. Cross-fall problem	11	0.7
36. Junctions or features too close	10	0.7
37. Possible overtaking problem	10	0.7
38. Other	40	2.6
Total safety problems	*1517*	*96.5*
Lack of information to carry out an audit of a feature (drawings or work not finished)	33	2.2
Inconsistency in design/layout/marking/signing	15	1.0
Inconsistency between drawings	4	
Acceptance for departures from standards has not been indicated	2	
Total insufficient information	*54*	*3.5*
All comments	*1571*	*100*

Safety audit problems

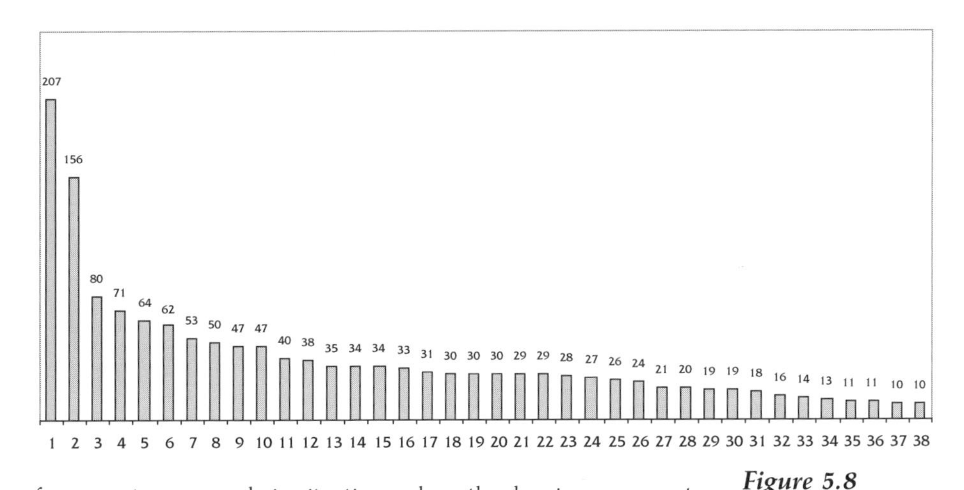

Figure 5.8
The most common issues identified – see Table 5.1 for key to problem numbers

of comments were made in situations where the drawings were not complete, or the scheme was not yet constructed when the Safety Auditors attempted to carry out a Stage 3 audit.

The most frequently referred to problems dealt with the inadequacy of road signs and markings. A more detailed explanation of the results from Table 5.1 and Fig. 5.8 is given below.

Inadequate road signs

- More likely to be a Stage 3 comment.
- In 71% of cases the appropriate signs were not provided.
- Poor signage for cyclists is over-represented.
- Poor signage for roundabouts and mini-roundabouts is over-represented.

Figure 5.9
A group of Safety Auditors examining the signs shown in Fig. 5.10

Figure 5.10
No wonder they were excited! The tourist sign has been placed directly in front of the roundabout direction sign

Inadequate road markings or road studs

- More likely to be a Stage 3 comment.
- In 74% of cases the appropriate markings were not provided.
- Poor road markings for cyclists are over-represented.

Figure 5.11
This cycle lane extends beyond the vehicle stop line, but there is no advanced stop line

Visibility to signs or traffic signals restricted

- More likely to be a Stage 3 comment.
- In 28% of cases visibility was restricted by vegetation.
- In 25% of cases visibility was restricted by poor sight lines or alignment.
- In 16% of cases visibility was restricted by other signs.

Figure 5.12
This direction sign is partly obscured by a large tree

- In 14% of cases visibility was restricted by a vehicle, for example, at a bus stop.
- Two-thirds of comments relate to junctions.
- Visibility at traffic signals is over-represented.

Inadequate tactile paving

- More likely to be a Stage 3 comment.
- Nearly half of all comments related to junctions.

Figure 5.13
There is no tactile paving stem at these signals, leading to the blind pedestrian missing the crossing

Problems with traffic lanes – number or width

- More likely to be a Stage 1 comment.
- Problems with cycle lanes are over-represented.
- Lanes at roundabouts or mini-roundabouts are over-represented.

Figure 5.14
The cycle lane is less than 0.7 m wide – inadequate to protect a cyclist

Inadequate lighting or reflectivity of signs

- More likely to be a Stage 3 comment.
- 16% of comments related to pedestrian crossings.
- Lighting for pedestrians is over-represented.
- Poor lighting at roundabouts or mini-roundabouts is over-represented.

Figure 5.15
Poor lighting is a safety issue at some roundabouts

Unsafe crossing point for vulnerable road users

- Crossing facilities at traffic signals are over-represented.
- 10% of comments related to 'wrong-way' staggers at pedestrian facilities.

Figure 5.16
Pedestrians cross through the central reserve of this dual carriageway to access a park – in conflict with high speed traffic

Problem for future maintenance

- More likely to be a Stage 3 comment.

Figure 5.17
Maintenance issues need to be considered, especially at complex signal junctions

Inadequate provision for vulnerable road users

- More likely to be a Stage 1 or 2 comment.
- Cyclists are over-represented.
- Poor provision at traffic signals is over-represented.
- Poor provision at roundabouts and mini-roundabouts is over-represented.

Figure 5.18
Confusion at the end of a cycle route – where should I go next?

Inadequate safety fence

- More likely to be a Stage 3 comment.
- Mainly a link related problem.
- Nearly half of all comments related to absence of safety fence.

Figure 5.19
Poles being constructed on the wrong side of the safety fence

Dropped crossing absent or not flush

- More likely to be a Stage 2 comment.
- Nearly half of all comments related to junctions (Fig. 5.20).

Poor design or absence of splitter island or refuge

- More likely to be a Stage 1 comment.
- Poor provision for pedestrians is over-represented.
- Mainly a junction problem (Fig. 5.21).

Figure 5.20
This elderly pedestrian is struggling with a 'dropped' kerb around 100 mm high

Figure 5.21
Pedestrian crossing a wide bellmouth at a road junction

Risk of actual speed being greater than design speed

- More likely to be a Stage 1 comment.
- Mainly a link related problem.

Figure 5.22
This 40 mph road appears much faster, and is driven at speeds well in excess of the speed limit

Stopping problem – absence or insufficient length of anti-skid

- More likely to be a Stage 3 comment.
- Surfacing at traffic signals is over-represented.
- Surfacing at roundabouts and mini-roundabouts is over-represented.

Figure 5.23
The provision of anti-skid road surfacing has been omitted – with the possibility of shunt accidents taking place at this crossing

Footway obstructed by poorly-sited street furniture

- More likely to be a Stage 3 comment.
- Mainly a link related problem.

Figure 5.24
This pole effectively blocks the footway

Visibility to pedestrians restricted through poor guard rail or planting

- More likely to be a Stage 2 comment.
- Nearly half of all comments related to junctions.

Figure 5.25
Conventional guard rail masked pedestrians –

Figure 5.26
– and was replaced by high visibility rails

Lack of continuity of a facility

- Mainly a cyclist related problem.
- Mainly a link related problem.

Figure 5.27
This cycle route is severed by the fencing

Obsolete road signs or markings

- More likely to be a Stage 3 comment.
- Half of all comments related to junctions.

Figure 5.28
The old sign at this location has not been removed, and it blocks the new sign

Specific highway feature inconspicuous

- 30% of features are priority junctions.
- 30% of features are roundabouts.

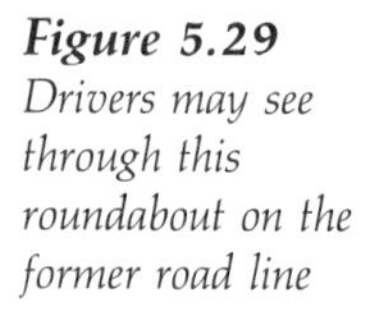

Figure 5.29
Drivers may see through this roundabout on the former road line

Forward visibility restricted

- More likely to be a Stage 1 comment.
- One-third of comments relate to priority junctions.

Figure 5.30
Forward visibility around the bend is restricted by the left-hand bank –

Figure 5.31
– leaving the junction obscured

- One-fifth of comments relate to roundabouts.

Set back from kerb edge too small

- More likely to be a Stage 2 comment.
- Set back at roundabouts and mini-roundabouts are over-represented.

Figure 5.32
Bollards placed close to the kerb edge pose a threat in a loss of control collision

Drainage problems and location of gullies

- More likely to be a Stage 3 comment.
- Pedestrians are over-represented.
- Cyclists are over-represented.

Figure 5.33
This gully could be a trip hazard for pedestrians

Horizontal alignment too severe

- More likely to be a Stage 1 or 2 comment.
- Roundabouts and mini-roundabouts are over-represented.

Figure 5.34
This roundabout is sited to the left of an approaching driver's field of vision

'See through' problem

- This refers to either pedestrians or motorists 'seeing through' to a misleading traffic signal.

Figure 5.35
Pedestrians may be encouraged to cross against a red signal by the far side 'green man'

- More likely to be a Stage 3 comment.
- Roundabouts and mini-roundabouts are over-represented.

Pedestrian guard rail absent or insufficient

- More likely to be a Stage 2 comment.
- Traffic signals and priority junctions are over-represented.

Figure 5.36
The hole in the guard rail may encourage some pedestrians to cross this busy dual carriageway

Absence of pedestrian crossing points

- More likely to be a Stage 1 or 2 comment.
- Roundabouts and mini-roundabouts are over-represented.

Figure 5.37
Pedestrian crossings can be successfully sited close to roundabouts to improve pedestrian access and safety

Poorly-sited street furniture

- More likely to be a Stage 2 comment.
- Mainly a problem for motorists – view obstructed in half of cases.
- Roundabouts and mini-roundabouts are over-represented.
- Traffic signals are over-represented.

Figure 5.38
Spot the traffic signals! The poles are completely obscured by trees

Type of junction control

- More likely to be a Stage 1 comment.

Figure 5.39
A mini-roundabout may not always be the safest form of junction control

Conflict between different road users

- More likely to be a Stage 2 comment.
- Mainly conflict between pedestrians and cyclists.
- Mainly at junctions.

Figure 5.40
Cyclists using the footway are perceived as a threat by pedestrians, particularly those with sight impairment

Poorly-designed traffic calming feature

- More likely to be a Stage 2 comment.
- Mainly related to motorists not slowing down enough.

Figure 5.41
Additional markings make this speed cushion appear more severe, but cushion spacing is an important issue in terms of travel speed

Poor design of bus stops

- More likely to be a Stage 1 comment.

Figure 5.42
Bus stop sign sited on a pole right at the kerb edge

Previous accident history unaccounted for

- More likely to be a Stage 1 or 2 comment.
- Nearly two-thirds of overlooked accidents were shunts.
- Nearly one-third were dark accidents.

Figure 5.43
Historical accident records should be checked for on-line improvement schemes

Figure 5.44
Vehicle 'launched' by safety fence end treatment

Loss of control hazard

- More likely to be a Stage 1 comment.

Parking problem

- More likely to be a Stage 3 comment.
- Mainly motorists on links.

Figure 5.45
Parked vehicle blocks cycle lane

Cross-fall problem

- More likely to be a Stage 3 comment.
- Two-thirds of comments were related to junctions.

Figure 5.46
Designing for drainage on roundabouts can lead to large vehicles overturning

Possible overtaking problem

- More likely to be a Stage 1 or 2 comment.
- Mainly motorists on links.

Figure 5.47 Single carriageway roads with the appearance of dual carriageways can lead to head-on collisions

Specific problems by stage of audit

Tables 5.2 to 5.5 show those problems most likely to be identified at each of the stages of audits. In each case the table lists those problems that have been identified at more than expected levels, when compared to the distribution of comments by audit stage shown in Fig. 5.5.

Table 5.2 *Audit comments made at Stage 1*

Safety Audit comment	*% of these comments made at Stage 1*
Junctions or features too close	60
Previous accident history unaccounted for	57
Poor design of bus stops	50
Junction control	42
Loss of control hazard	38
Poor design or absence of splitter island or refuge	37
Problems with traffic lanes – number or width	36
Risk of actual speed being greater than design speed	35
Forward visibility restricted	34
Inadequate provision for pedestrians or cyclists	34
Poorly-designed traffic calming feature	33
Unsafe crossing point for vulnerable road users	33
Parking problem	27
'See through' problem	27
Horizontal alignment too severe	26
Set back from kerb edge too small	24
Lack of continuity of a facility	20
Absence of pedestrian crossing points	19
Visibility to signs or traffic signals restricted	19
Cross-fall problem	18

Table 5.3 Audit comments made at Stage 1/2

Safety Audit comment	*% of these comments made at Stage 1/2*
Possible overtaking problem	50
Poorly sited street furniture	45
Visibility to pedestrians restricted through poor guard rail or planting	42
Pedestrian guard rail absent or inadequate	29
Absence of pedestrian crossing points	29
Horizontal alignment too severe	26
Poor design or absence of splitter island or refuge	23
Dropped crossing absent or not flush	21
Stopping problem – absence or insufficient length of anti-skid	21
Unsafe crossing point for vulnerable road users	19
Parking problem	18
Problems with traffic lanes – number or width	18
Set back from kerb edge too small	17
Inadequate road signs	16
Loss of control hazard	15
Footway obstructed by poorly sited street furniture	15

Stage 1

Around 18% of all the 1571 comments were made at Stage 1, but 60% of the references to junctions or features being too close were made at Stage 1.

Stage 1/2

14% of all the 1571 comments were made at Stage 1/2, but 50% of the references to possible overtaking problems were made at Stage 1/2.

Stage 2

23% of all the 1571 comments were made at Stage 2, but 53% of the references to conflict between road users were made at Stage 2.

Stage 3

44% of all the 1571 comments were made at Stage 3, but 92% of the references to problems for future maintenance were made at Stage 3.

Although the comments at each stage are not mutually exclusive, there is a considerable difference between those made at the early stages compared to later on. Only four comments – parking, 'see-through', restricted visibility to signs/signals and cross-fall – are made at above expected levels at both Stage 1 and Stage 3.

***Table 5.4** Audit comments made at Stage 2*

Safety Audit comment	*% of these comments made at Stage 2*
Conflict between different road users	53
Dropped crossing absent or not flush	47
Poorly-designed traffic calming feature	44
Junctions or features too close	40
Previous accident history unaccounted for	36
Set back from kerb edge too small	34
Inadequate provision for pedestrians or cyclists	34
Lack of continuity of a facility	33
Absence of pedestrian crossing points	33
Possible overtaking problem	30
Inadequate tactile paving	28
Drainage problems and location of gullies	28
Inadequate road signs	28

Most of the Stage 1 problems are fundamental design issues, while a lot of the Stage 3 problems relate to construction issues.

The work described above confirms and expands upon earlier findings from a study carried out by TMS Consultancy in conjunction with other Road Safety Auditors. In 1998, The Automobile Association

***Table 5.5** – Audit comments made at Stage 3*

Safety Audit comment	*% of these comments made at Stage 3*
Problem for future maintenance	92
Inadequate lighting/reflectivity	76
Inadequate safety fence	64
Specific feature inconspicuous	63
Obsolete road signs or markings	63
'See through' problem	62
Inadequate tactile paving	56
Stopping problem – absence or insufficient length of anti-skid	56
Visibility to signs or traffic signals restricted	55
Cross-fall problem	55
Inadequate road markings or road studs	52
Drainage problems and location of gullies	50
Inadequate road signs	49
Footway obstructed by poorly-sited street furniture	48
Parking problem	45

and TMS Consultancy brought together a group of seven experienced Road Safety Auditors from around the UK to produce a booklet to highlight some of the problems most commonly identified within Road Safety Audit.[1]

Reference

1. AA Policy/TMS Consultancy (1999). 'What goes wrong in Highway Design'

6. International experience

The Institute of Transport Engineers (ITE), a multinational organisation based in the US, produced an 'Information Report'[1] in 1994, summing up the current status of Road Safety Audit throughout the world. The ITE report confirmed that much of the Safety Audit expertise at the time was in the UK and Australasia. It noted that, although the process would be beneficial in North America, various problem areas – especially legal issues – should be addressed before it could be introduced in the US.

The ITE report noted that there was great scope for implementing Road Safety Audits in developing countries, especially if they were linked to the infrastructure investments being made by funding agencies such as the World Bank. This would appear to be an ideal area for technology transfer programmes to operate, since Safety Audit is as applicable to small-scale local-build schemes as it is to massive, outside funded highway developments. While working on a World Bank project in Peru in the mid-1990s, a set of Safety Audit procedures were drafted by TMS Consultancy.

Figure 6.1
Pedestrians in Peru ignore a footbridge and cross a high speed road at grade

Figure 6.2
Modern highway in Peru. The marker posts help to indicate the alignment in conditions of poor visibility

Figure 6.3
Poor siting of bus stops – encouraging pedestrians to cross this high speed road

Australia and New Zealand

Contribution by Phillip Jordan (Vic Roads, Australia) and Ian Appleton (Transfund New Zealand)

The concept of Safety Audit spread to Australia in the early 1990s. Senior Engineer Phil Jordan of VicRoads based in Melbourne, was seconded to UK Highway Authorities where the Safety Audit process was being developed. Within a short time other Australian states had also taken up the Safety Audit principle.

Safety Audit in New Zealand progressed in two stages. The initial work concerned the Safety Audit of new construction projects and was undertaken in 1991/1992 with workshops and pilot exercises on state highways with assistance from the UK and Australia.

A working party was set up with representatives from all sectors. This working party prepared the 'New Zealand Safety Audit Manual'.[2] In July 1993, Transit New Zealand adopted the manual and a policy to apply it to a 20% sample of state highway projects each year. Since that date, safety auditing on state highway projects has been an operational activity.

There has never been any requirement that Safety Audits be undertaken on local roads projects. However, about one-third of local authorities say they conduct Safety Audits, while another third, mainly small rural authorities, have no projects to audit. To promote Safety Audit in the local authority sector a series of pilot audits were conducted using teams of engineers from neighbouring authorities, led by a consultant.

The Safety Audit of existing roads has taken a different turn in New Zealand. This is because of the existence of a separate funding agency, Transfund New Zealand. The Safety Audit of existing roads has been developed by Transfund essentially as a tool to determine whether a road controlling authority is doing a good job in respect of road safety. These audits look for recurring patterns of deficiencies across an authority's network. This work was started in 1995 with a series of pilot audits. Procedures for these audits were finally published in 1998.

Safety auditing has been well accepted by the profession. There are some issues that need to be resolved, and no doubt there are improvements to be made. Currently there are debates over a number of improvements or enhancements to the Safety Audit process, including such matters as making Safety Audit compulsory, and the training and accreditation of Safety Auditors.

Transit New Zealand is a member of Austroads* and therefore participates in joint projects including Safety Audit projects. This has been of considerable benefit to New Zealand to the extent that Australian engineers are now employed on audit teams in New Zealand.

The outcome of this activity in New Zealand and Australia was a unique collaboration in producing the Austroads publication 'Road Safety Audit',[3] which gave clear guidelines applicable to operations in both nations and their component local highway authorities. This book is a definitive document on Safety Audit, for its messages and recommended procedures transcend hemispheres and are applicable anywhere in the world. The guidelines are currently being updated and are due to be re-issued in late 2001.

Austroads also ran the first International Conference on Safety Audit in Melbourne during May 1998 and brought together over 200 people who were working (or had an interest) in this subject.

* Association of State Road Authorities of Australia and New Zealand.

North America

Contribution by Frank Navin (University of British Columbia)

The first reported Road Safety Audit in Canada was a 50% design stage audit of a High Occupancy Vehicle (HOV) lane for Highway 1 in suburban Vancouver, carried out for the British Columbia Ministry of Transportation and Communications in 1997. In the US, the Pennsylvania Department of Transport (DOT) carried out extensive feasibility of Safety Audit starting in April 1997. The model followed by both the US and Canada evolved from the UK and Australian model.

These first experiences highlighted the fact that the audit process needed to be tailored to local conditions. The main departure from previous experience was the requirement by the Canadian client that a set of general safety recommendations be included. The Pennsylvania DOT gives verbal recommendations during the audit report handover.

Many of the Safety Audits carried out in British Columbia, Alberta and Pennsylvania are at the design stage. The typical recommendations, in order of frequency cited include:

- Improve bicycle and pedestrian facilities, and cross-section changes such as lane width;
- Improve sight distance, better provisions for trucks and buses, improve drainage, and improve interchange ramps;
- Improve signing, provide median barriers or end treatments, and improve weaving area.

In general, the findings indicate that designers typically address the horizontal and vertical alignment with a high level of built-in safety, but issues related to the cross-section such as lane width are more prone to safety concerns. Vulnerable road users such as cyclists and pedestrians are also a common safety concern.

The Pennsylvania DOT found similar problems during their Safety Audit pilot programme. In addition, they cannot make any changes affecting the alignment envelope after official acceptance of the Environmental Impact Assessment.

The term 'audit' has raised concerns from certain quarters of the legal profession in North America. The Ministry of Transportation in Ontario calls Safety Audit, 'Operational Assessments'. All of the procedures of the traditional Safety Audit are also part of the Ontario

Operational Assessment. Similar concerns have caused the Safety Auditors of the Fredericton-Moncton Highway in New Brunswick to include a disclaimer on the cover page that describes the limitations of the audit process. The Pennsylvania DOT audit report is a privileged document and is not discoverable in a trial.

Safety Audit is continuing to evolve in North America. In Canada, the Transportation Association of Canada has accepted a recommended Safety Audit procedure. The United States Federal Highway Administration sponsored a study tour to Australasia in the late 1990s and the Institute for Transportation Engineers in the US is coordinating some of the US-FHWA's Road Safety Audit education.

Asia

In South East Asia, Malaysia has brought into operation Safety Audit procedures based upon a very comprehensive guidelines document, and Singapore is also starting to bring Safety Audits into action as it develops and refines its road system. Hong Kong is also taking the first steps to develop a Safety Audit procedure of its own.

Europe

Outside of the United Kingdom the progress of formal Safety Audit in the rest of Europe has been relatively slow with the exception of

Figure 6.4
In the Netherlands, a well developed hierarchy of roads and separation of road users helps to minimise conflict

Denmark and Ireland. The Danish Government's Road Directorate has developed and implemented a Safety Audit process[4]. It is based upon the UK model and the Directorate has produced comprehensive documentation on the subject.

In Ireland a safety engineering manual written for the government by TMS Consultancy was published in 1996 and included the concept of safety checking road schemes. The National Roads Authority, assisted by TMS, has recently introduced comprehensive Safety Audit procedures[5].

Elsewhere in Europe, Iceland is currently carrying out some Road Safety Audit work. Safety engineers in Germany, France, Greece, Italy, Portugal and Sweden are also interested in the audit process and are looking at the potential for developing and implementing procedures as well as carrying out training and undertaking trial audits. In Italy, the University of Naples has been at the forefront of research in this area.

Figure 6.5
Narrow bridge on the Iceland Ring Road – signs and markings have been installed to reduce accident potential

In 1997 the European Transport Safety Council (ETSC) recommended that Member States should introduce a mandatory requirement that all new major road schemes should be subjected to an independent Road Safety Audit. ETSC went on to advise that in time these formal procedures should be extended to smaller schemes and existing roads[6].

Figure 6.6
Shared pedestrian/ cycle path in Iceland – under-used by both users in these conditions!

References

1. ITE (1994). 'Information Report'
2. Transit New Zealand (1993). 'New Zealand Road Safety Audit Manual'
3. Austroads (1994). 'Road Safety Audit'
4. Ministry of Transport, Denmark (1997). 'Manual of Road Safety Audit, Road Directorate'
5. HD 19/00 (2000.) 'Road Safety Audits', National Roads Authority
6. ETSC (1997). 'Road Safety Audit and Road Safety Impact Assessment'

7. Legal implications of the Road Safety Audit process

Following a road accident it is possible that a victim or other claimant (for example, someone who has had a claim made against them) will make allegations against other parties involved in the accident. Occasionally highway authorities, design consultants, and safety auditors can become subject to specific allegations of breach of statutory duty or negligence. Although it is possible for criminal proceedings to be brought against those involved in Safety Audit, it is more likely that a claimant will sue in a civil court for a breach of statutory duty and/or negligence.

The comments made in this chapter relate to UK law, but some of the principles can be applied in other countries too.

In order to find against a defendant in a criminal case, the evidence presented to the court must be proved beyond all reasonable doubt. The burden of proof in a civil court is different – the case can be won on the balance of probability. There are many statutory obligations that highway and local authorities have with respect to highways, but two of the most important are noted below.

Highways Act (41)[1]

The 1980 Highways Act states that: 'the authority who are, for the time being, the highways authority for a highway maintainable at the public expense, are under a duty... to maintain the highway'. This places a statutory duty of care on highway authorities to maintain public highways. In addition, local authorities have statutory road safety duties.

Road Traffic Act (39)[2]

The 1988 Road Traffic Act, states that each local authority: '... in constructing new roads, must take such measures as appear... to be appropriate to reduce the possibilities of such accidents when the roads come into use'. Some solicitors working on behalf of road accident victims may interpret this as a legal obligation to carry out Safety Audits, particularly on new roads, many of which are being constructed under developer-led funding. Indeed, Safety Audits are required on trunk roads and IHT guidance recommends the undertaking of Safety Audits as 'best practice'.

One definition of negligence is 'someone who does foreseeable harm to someone else' (Lord Aitken). A Safety Auditor looks into the future when carrying out a Safety Audit and tries to assess future risk to road users (foreseeing harm). Because of this, a view has developed that undertaking Safety Audit actually increases the possibility of litigation after an accident has occurred on a new road scheme.

A contrasting view suggests that carrying out Safety Audit in line with established procedures reduces the chances of being found liable. This is because reasonable and competent efforts will have been made to foresee and prevent harm, and because safety will be shown to have been added to a scheme through the Safety Audit process.

Highway authorities defending a claim can refer to the Highways Act in preparation of their defence.

Highways Act (58)[3]

The Highways Act states that 'It is a defence ... to prove that the highway authority had taken such care as in all the circumstances was reasonably required to ensure that the highway was not dangerous for traffic'. After an accident has occurred, it may be alleged that a highway issue is in part responsible for the accident. If the accident has occurred on a road scheme that should have been audited, the Safety Audit report, exception report and any other information are discoverable documents that may have a bearing on the outcome.

Claimant's success

The claimant's success depends on that person being able to show that on the balance of probability, reasonable care has not been taken on this occasion. In order to defend a claim, a local authority or consultant may have to demonstrate that its formal Safety Audit procedures comply with industry standards, and that on this occasion it followed the procedures in a reasonable manner.

The Safety Auditors' dilemma

The Safety Auditors' main dilemma is that they might fail to spot something that later is shown to be a factor in an accident. The claimant

Figure 7.1
This car hit a tree – but was there a highway issue that should have been spotted by a Safety Auditor?

may try to prove that the 'reasonable auditor' would have identified the problem. The Safety Auditors are then likely to find themselves in one of the following situations.

- The safety problem was discussed but not submitted in the Safety Audit report. It may have been submitted as a problem at a previous stage of Safety Audit and rejected in a corresponding exception report.
- The safety problem affected part of the scheme that was considered to be outside the Safety Audit brief.
- Road safety knowledge has changed since the Safety Audit was carried out. At the time of the accident it would have been unreasonable to foresee that type of problem.
- The 'safety' problem was considered, but thought to be a 'non-safety' issue by the Safety Auditors.

In case any of these scenarios develop, Safety Auditors are advised to maintain good records of their Safety Audit process, including any checklists used during the process, and to spell out precisely what information has been used for Safety Audit purposes. In the case of receiving an exception report on a problem at an early stage of the Safety Audit, the Safety Auditors are advised to repeat the road safety problem at subsequent stages if they feel it is still appropriate.

However, it may be that none of these scenarios apply, in which case the Safety Auditors may have simply made a bad mistake, and overlooked a potential accident problem that other Safety Auditors would have been expected to identify. In this case there is a possibility of liability against the Safety Auditors.

The project manager or designers' dilemma

In this scenario, the situation may occur in which the Safety Auditor makes a comment, but the recommendation is overturned within the exception report. The project manager or designer may have to satisfy the court that their reason for rejecting the Audit recommendation was 'reasonable'.

Alternatively the Safety Auditor may describe a road safety problem, but the recommendation is ignored and there is no exception report. The person responding to the Safety Audit has a potential liability problem.

The client's dilemma

Following an accident involving an allegation of highway liability, the client will be seen to be responsible for managing the entire design process, including Safety Audit. The client should be concerned in a situation where there is no Safety Audit at all, or if it fails to appoint a suitably qualified Safety Audit Team, or where there is no exception report.

Concluding remarks on legal implications

Safety Auditors should be aware that the language used in the audit report is important. There is a big difference between stating that 'deflection *must* be applied to this approach to the roundabout', and stating that 'consideration should be given' to applying deflection.

Individual Safety Auditors should remember that ownership for the scheme remains with the designer and/or client.

While the objective of Safety Audit is to improve road safety, Safety Auditors should be aware of the legal implications of their task. Those involved in drawing up procedures for Safety Audit, are advised to obtain clear management commitment for the task and to take legal advice on the procedures.

References

1. Highways Act 1980, section 41. HMSO
2. Road Traffic Act 1988, section 39. HMSO
3. Highways Act 1980, section 58. HMSO

8. Conclusions

Costs and benefits of Safety Audit

There could be concern that Safety Audit adds unnecessarily to the cost of a scheme and lengthens the time for the scheme to proceed towards implementation. On the other hand, highway authorities have spent millions of pounds on remedial treatments to road schemes less than five years old, resulting in inconvenience, delay, and added risk to millions of road users. So intuitively, prevention is better than cure.

Costs

The average cost of each stage of a Safety Audit in the UK is estimated at between £500 and £1000 (although this could be considerably more for a few complex schemes, such as a new section of motorway or a complex set of linked urban traffic signals). This gives a total audit cost on a scheme for Stages 1 to 3 of between £1500 and £3000. This figure should be compared with the average cost of an injury accident in the UK of some £70 000, and some £1.3 million for a fatality.[1]

There are a number of costs that can be attributed to a Safety Audit. First there is the cost of the audit itself. The cost of a Safety Audit is related to the time spent to complete it, rather than the cost of the scheme itself. It takes less time to audit a scheme involving a new link road with a simple junction at each end than it does to audit a complex traffic signal junction in an urban area. Research carried out by the IHT found that the average time taken to complete an audit was 25 hours.[2]

The second element of cost relates to the implementation of the recommendations contained within the audit report. In general, these costs are not significantly high and items identified at Feasibility Stage and Stages 1 and 2 may have no cost implications at all (although they may require some redesign time). There are, however, some instances where audit recommendations, particularly at Stage 3, will add to the cost of a scheme. For example, a recommendation for applying anti-skid surfacing on the approach to a set of traffic signals.

The implementation cost of audit recommendations will obviously vary greatly between schemes, but an average figure of £5000 per scheme would seem to be a reasonable assumption. The IHT research referred to earlier indicated that about half of Safety Audits involve redesign and that increases in design costs were in the order of 1% of overall scheme costs.

Adding these costs together (assuming an audit is carried out at three stages) would give costs in the order of £10 000 per scheme, although a Safety Audit should be seen as part of the scheme design process and cost, rather than additional.

Benefits

It is difficult to identify the benefits of carrying out a Safety Audit on a scheme in a quantitative way. When an audit has been carried out, the scenarios are that either the recommendations are implemented or they are not; and although the subsequent accident record can be examined, only one of the scenarios can be evaluated. It is not possible to compare how an individual scheme that has been audited could have performed had the audit not been carried out.

The only monitoring of Safety Audited schemes in the UK seems to have been carried out by Surrey County Council.[3] The County Council was introducing Safety Audit at the time and they checked the safety performance of 20 minor improvement schemes that had been audited and modified accordingly, against 20 similar schemes that had not been audited. Their findings were that the audited schemes had, on average, about one casualty per year less than the non-audited schemes.

It may be possible to look at schemes in a wider national context and check the performance of audited schemes versus those that have not been audited. This is difficult within a local authority as most authorities either audit schemes of a certain type or they do not. It would be inadvisable not to carry out Safety Audits as part of an experiment, due to potential legal liabilities.

One way of estimating whether Safety Audit could prevent accidents is to carry out an audit on a road that has been open for, say, five years, and then compare the Safety Audit comments with the accident record. A TMS Consultancy Audit of an existing road in Ireland showed that 30% of identified audit problems materialised as injury accidents. The implication is that some of these accidents could have been prevented if the Safety Audit had been carried out at the design stages.

Some work carried out in New Zealand suggests that the benefit to cost ratio for Safety Audits is in the order of twenty to one. In Denmark,[4] the first year rate of return for Safety Audits has been estimated as over 149%. This figure was based on estimates for accident savings that might be made by introducing Safety Audit recommendations.

There are also cost savings to be made by making changes to a scheme during its design rather than after construction (and possibly after accidents had occurred). The TRL studied 22 schemes and estimated that if design changes were made at Stage 1 or Stage 2 rather than after completion, a saving of £11 000 per audit would be achieved[5].

Figure 8.1
It is much cheaper to amend schemes during design – compared with making changes once the scheme is complete

A further qualitative benefit is the extent to which design engineers receive improved safety awareness through the Safety Audit process. Local authorities in the UK who have carried out this work over a decade or more have noticed a reduction in the number of comments being made by their Safety Auditors.

Using current accident costs, Safety Audit costs identified in this chapter, and assuming the safety benefits identified by Surrey County Council, a first year rate of return of around 600% can be estimated for this type of work.

Acceptance and rejection of Safety Audit comments

The success of a Safety Audit can be measured not only by cost/ benefit analysis, but also by the proportion of problems and recommendations that are accepted by a client. Safety Audit recommendations should be relative to the problem and the stage of the design to ensure that a high percentage of comments are not rejected.

In Denmark, interviews during a Safety Audit pilot study[4] showed that most clients were satisfied with the level of audit comment. In the UK, a study 're-auditing' a number of schemes showed good consistency between audit comments from the different sets of Safety Auditors, hinting at 'responsible' audit comments.[5]

It is good practice to send out a feedback form with Safety Audits with the objective of getting clients to tell the Safety Auditors how many of the safety problems are accepted, and how many of the recommendations will be implemented. From the feedback forms returned to TMS Consultancy, an average of around 60–70% of audit recommendations are implemented.

The SaferCity Project in Gloucester has produced exception reports in response to all Safety Audit reports produced by TMS. Overall, 73% of the Safety Audit 'problems' were accepted, with SaferCity either fully accepting the recommendation, or implementing an alternative solution.

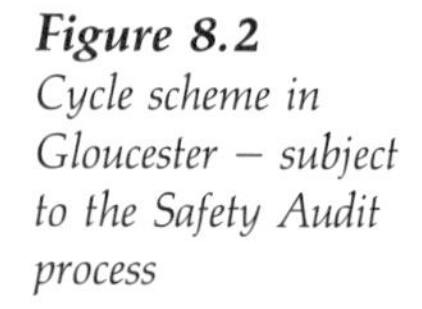

Figure 8.2
Cycle scheme in Gloucester – subject to the Safety Audit process

In some cases a client is unable to adopt recommendations due to design constraints. In other cases the client does not believe that the

Table 8.1 *Safety issues most frequently accepted by the client*

Description	*% accepted*
Obsolete road signing	100
Visibility of pedestrians restricted	100
Problem for future maintenance	92
Poor design of build-outs and/or bus stops	83
View obstructed by poorly-sited street furniture	80
Pedestrian route obstructed	76
Location of crossing might cause confusion	75
Absence of pedestrian crossing points	75
Colour of road surface missing or wrongly used	75
Inadequate road markings	71
Illumination or reflection of signs, bollards and other street furniture	67
Visibility of signs or signals restricted	67
Inadequate tactile paving	67
Feature inconspicuous	60
Poor design or absence of splitter island or refuge	57
Inadequate signing	54

issue raised is a real safety problem that could lead to road accident casualties. Tables 8.1 and 8.2 show the safety issues most frequently accepted and rejected by SaferCity in response to TMS Safety Audits.

Table 8.2 lists the issues most frequently rejected by the client, in descending order. In each case the Safety Auditors raised the issue on at least three occasions.

The list in Table 8.2 reflects those items that are difficult to fix within the constraints of the scheme, for example, continuity of cycle route, and the Safety Auditors' perception that more traffic calming may be required. It also reflects those items that the client feels are not safety issues, for example, the implementation of anti-skid at an improved set of signals or pedestrian facility where there is no record of existing wet skid accidents.

Table 8.2 *Safety issues most frequently rejected by the client*

Description	*% rejected*
Stopping problem. Absence or insufficient length of anti-skid	75
Conflicts between different road users	58
Poor continuity of cycle provision	57
Dropped crossings absent or not flush	54
Speed cushions or humps may not be effective	54

The future development of Safety Audit

A number of issues are currently being examined by Safety Auditors, clients and designers. This section of the book examines some of the more controversial issues that need to be addressed in the near future, and which may lead to changes in Safety Audit procedures in the UK.

Which schemes should be audited?

At present, Road Safety Audit is mandatory on all new trunk road schemes in the UK, and the majority of highway authorities carry out Safety Audits on local road schemes. It may be argued that Safety Audits should be mandatory for all road schemes.

However, there is clearly a cost implication in carrying out this work. Is it cost-effective to insist on a Safety Audit if the Audit turns out to be a substantial part of the overall cost? It could be argued that there should be a cut-off based on the cost of the scheme. But this book has shown that some safety issues are related to very small details on schemes, and a cost cut-off of, say, £10 000 would miss some schemes. Safety Audit is not an add-on – it is part of the design process.

The DETR Road Safety Strategy[6] requires local authorities to carry out child road safety audits – although there is little or no guidance within the strategy as to what these should be. Some local authorities have interpreted the requirement at a strategic level – looking at all accidents involving children and drawing up an action plan, while others have stated that they will add a specific section to their Safety Audit reports to look at problems and recommendations affecting child road safety.

The auditing of development-led road schemes within the planning process is another area that needs clarification.

At present, many local authorities require a Safety Audit too late in the process to address fundamental safety issues. Once planning approval has been given it is difficult to require developers to make significant changes to schemes, especially if they are costly or reduce the amount of land available for development. A flow chart illustrating the potential complexity of this process is shown at Fig. 8.4.

If the developer submits a Stage F Safety Audit with their planning application, the Safety Audit and response could be

Figure 8.3
Many internal layouts at development sites are not audited – here the absence of tactile paving could lead to conflict

considered before planning permission is granted. In a few cases, the findings of the Safety Audit could lead to the refusal of the planning application. It is more likely that the developer would be given planning permission, but that the scheme proceeds subject to specific requirements resulting from the Stage F Safety Audit. Subsequent Safety Audits at Stages 1, 2 and 3 could be enforced through Highways Act, section 38 or 278 agreements between the highway authority and the developer. In order to facilitate such a process, it would be necessary for senior highways and planning colleagues to review their in-house planning application processes, so that the inclusion of a Stage F Safety Audit becomes a requirement of the planning application.

What qualifications should Safety Auditors have?

A Safety Audit Team should comprise at least two people who are independent of the design team. This independence is vital to ensure that the design team does not influence the recommendations of the Safety Audit and therefore compromise safety at the expense of another issue. Team members should have recent relevant experience of undertaking Safety Audits and should also have more general road safety engineering experience.

Training of Road Safety Auditors is essential and any Audit Team member should have attended recognised road safety engineering training and Safety Audit training courses.

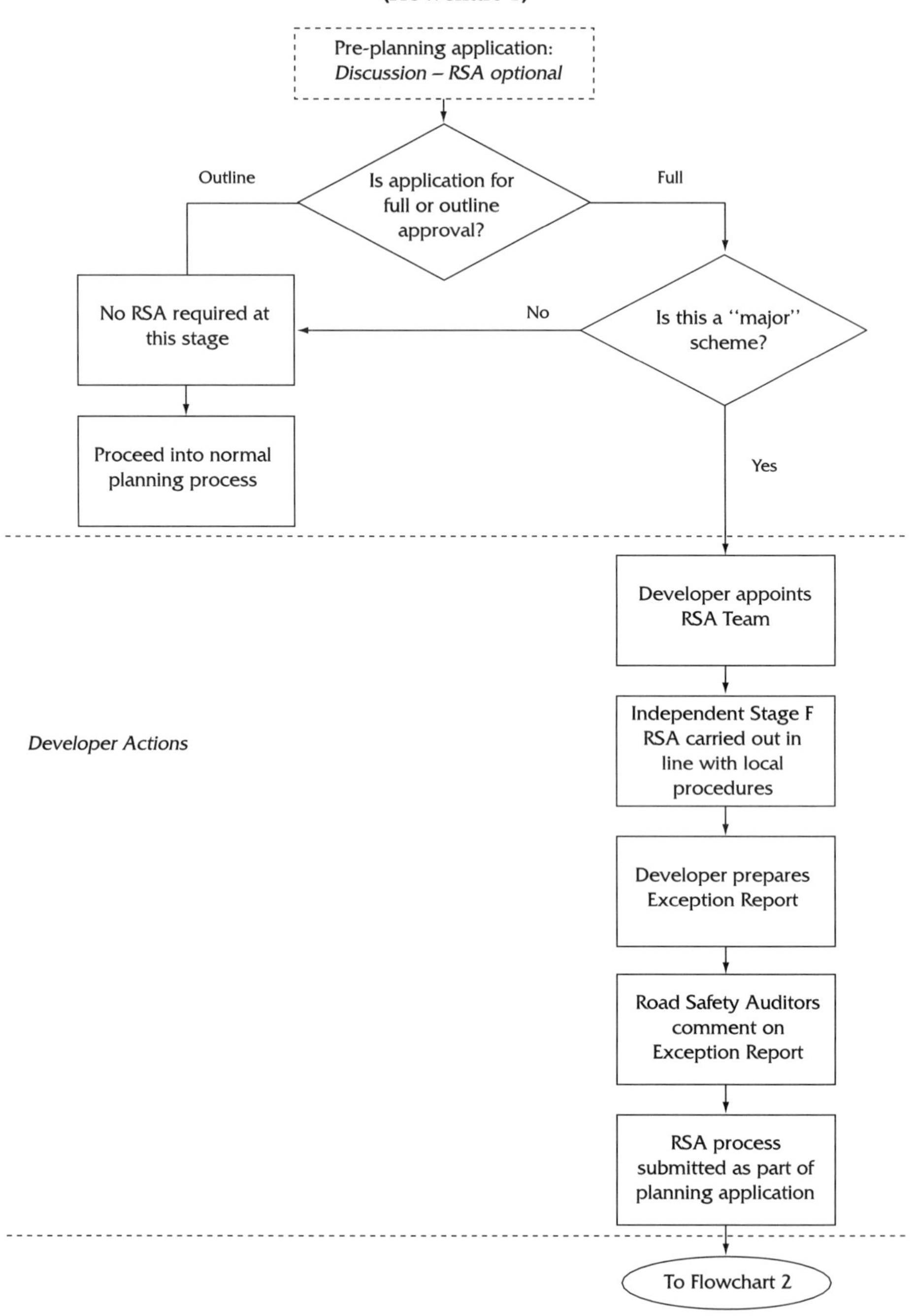

Figure 8.4 *Flowchart 1*

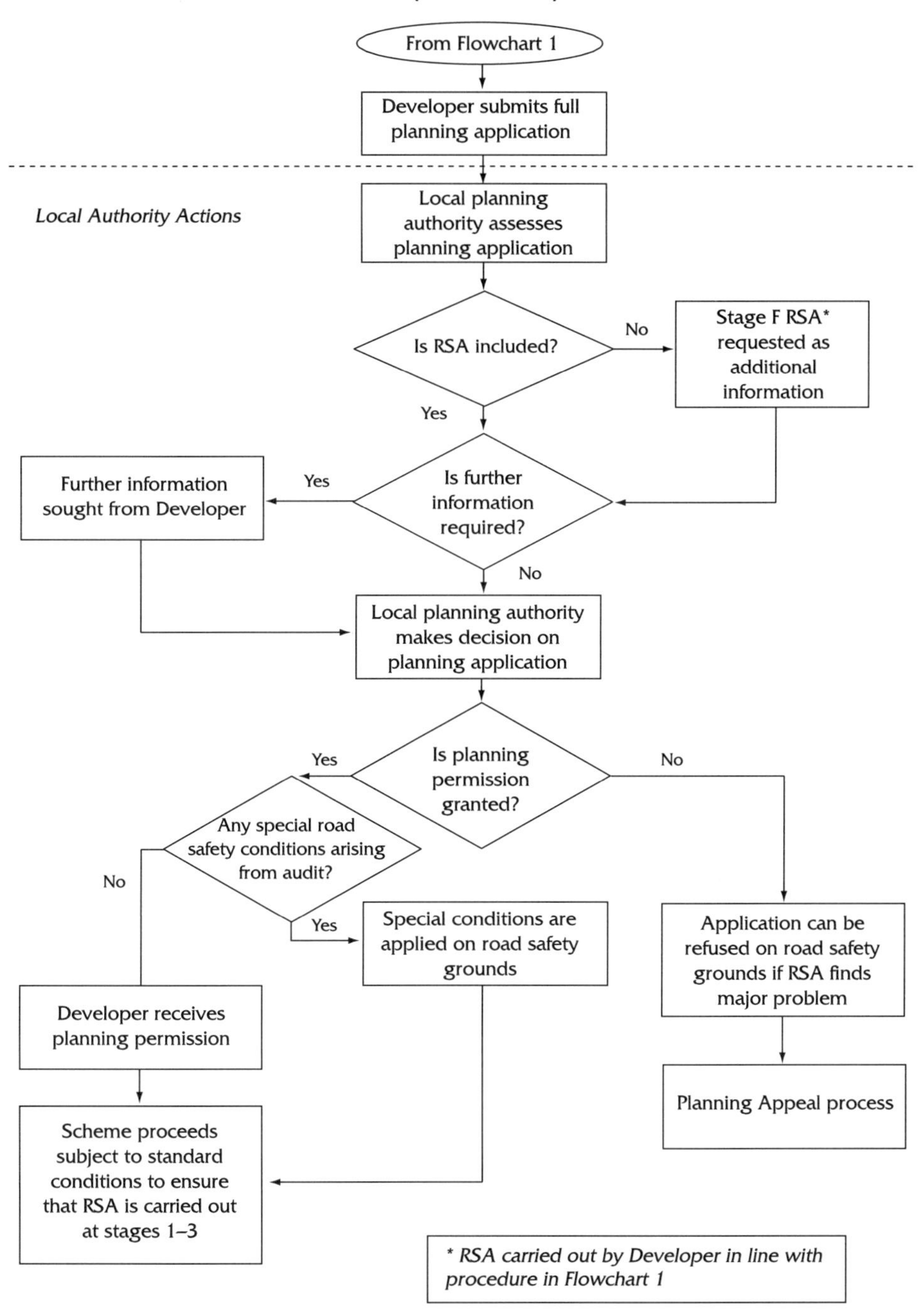

Figure 8.4 *Flowchart 2*

Figure 8.5
Delegates on a recent TMS Safety Audit course

The problem at present is that there is no formal qualification in Safety Audit, and it is possible for staff with very little safety engineering experience to produce audits that are at best a check on design standards.

A system of accreditation for Safety Auditors has been suggested, but there are some unresolved issues relating to the type of training courses that are appropriate, and the amount of experience necessary to become a team member, and then a team leader. One way forward is for organisations to appoint staff with an appropriate background to be Safety Audit Team observers – a third member of the team who does not sign the report. After a series of 10–15 Safety Audits have been carried out, that person may be acceptable as a team member.

The IHT is currently promoting an NVQ in Road Safety – which includes a full module on Road Safety Audit. Obtaining this module is a good way for Safety Auditors to demonstrate competence in this area and obtain a formal qualification.

TMS Consultancy has advised governments in situations where there is very little background safety engineering experience. In these situations, the following method for training up qualified Safety Audit teams has been recommended.

Qualifications for Safety Auditors

The following are some suggestions for establishing qualifications for staff undertaking Road Safety Audits.

Staff carrying out Safety Audits are likely to fall into four categories

A. experience in accident analysis;
B. experience in road design;
C. experience in traffic engineering;
D. newly-qualified engineers.

The four categories will all need some training in Safety Audit, but the level of training will be different. The following training is suggested for each of the groups:

A. short training course (five days) in Safety Audit theory and practice;
B./C. two-week course (five days accident analysis/five days Safety Audit);
D. one-year course in safety engineering including module in Safety Audit.

After formal training it is important that Safety Auditors gain practical experience in carrying out Safety Audits before they become Safety Audit Team leaders. The suggested experience before becoming a team leader is as follows.

A./B./C. Third person on five audits (audits on training courses can be included) and then second person on ten audits.
D. Third person on ten audits and then second person on twenty audits.

If possible, the audits carried out in this training period should include all stages of audit and different types of scheme.

Feedback loops – audit and exception reports

It is important to ensure an adequate response from the design team via the exception report process. The current standard for Road Safety Audit requires the production of exception reports, but it is not known what proportion of these are formally completed. Safety Auditors working on Highways Agency schemes are not supplied with exception reports as a matter of course.

It would be beneficial to provide more feedback and indeed greater interaction between Safety Auditors and designers throughout the process. A meeting between designers and Safety Auditors at the start and end of the audit process would be useful, and Safety Auditors could be asked to provide advice on safety issues during the design. However the independence of the Audit Team is vital and should not be compromised. The danger of a more open process is that designers and Safety Auditors working more closely together would lead to a less critical audit report, particularly where costly or politically difficult recommendations are required to overcome fundamental safety concerns.

Audits on existing roads

In many countries the Safety Audit process is being extended to carry out Safety Audits on existing roads, with no proposals for improvement at the time of the audit. In some of these countries the quality of accident data is poor, and reporting levels are low. The Safety Audit report is therefore used as a proxy for accident data.

If this type of work takes place in the UK there is a concern about the relationship with accident reduction work, particularly on routes. Accident reduction work has produced significant casualty reductions and excellent cost/benefit returns at single sites and along routes in the UK for the past 35 years. Accident reduction schemes are based on an analysis of historical accident records, and the definition of problems arising from those records. It would be difficult to justify spending large sums of money to fix problems identified from a Safety Audit if there were no or few recorded accidents.

It is interesting to note that Australian Road Safety Auditors now seem to be placing a reduced emphasis on this type of work, after pioneering this type of approach in the early 1990s, and Safety Auditors in North America have also reduced their interest in Stage 5 Audits.

Risk assessment in Road Safety Audit

For some time a view has been developing that the Safety Audit is 'power without responsibility'. The Auditor can make comments on safety issues and make recommendations, but has no direct ownership of the scheme (rightly so, maintaining independence). However, there are two concerns:

- Safety Auditors with limited safety experience make recommendations that change schemes without any safety benefits to be gained.

- Safety Auditors make recommendations where the cost of implementation far outweighs any safety benefit to be gained.

It has therefore been suggested that Safety Auditors carry out a formal risk assessment of their work, ranking both the audit problems and recommendations using a matrix like that shown in Table 8.3.

Table 8.3 *Risk assessment of Safety Audit*

Probability of outcome → **Severity of outcome ↓**	*More than once per year (probable score 4)*	*Once every 1–3 years (possible score 3)*	*Once every 3–7 years (remote score 2)*	*Once every 7–20 years (improbable score 1)*
Multiple fatal *(extreme score 4)*	16	12	8	4
Fatal/serious *(severe score 3)*	12	9	6	3
Minor injury *(minor score 2)*	8	6	4	2
Damage only *(negligible score 1)*	4	3	2	1

A risk score of 1–3 is 'low' risk, 4–9 is 'medium' risk, and 12–16 is 'high' risk. The Auditor would go through the report and give each problem a risk score – effectively their assessment of risk if nothing is done. The Auditor would then go back through their recommendations, and, making the assumption that the recommendation will be carried out, the Auditor re-assesses the risk.

An audit report could then contain a risk assessment table in the summary (see Table 8.4).

The information would be used by the client to help decide whether or not to implement the recommendations. The client could instruct the designer to cost the recommendations and then judge whether the

Table 8.4 *Risk assessment table*

Audit paragraph number	*Risk assessment of problem*	*Risk assessment if recommendation implemented*
2.1	16 (high)	9 (medium)
2.2	8 (medium)	1 (low)
2.3	6 (medium)	1 (low)
2.4	3 (low)	1 (low)
2.5	2 (low)	1 (low)

reduction in risk was worth the cost of improvement. At present it is sometimes too easy for a client or designer to turn down a Safety Audit recommendation on the basis of cost – this is not always reasonable.

Concluding remarks

This book has examined the Safety Audit process, provided extensive examples of road safety control data, and investigated the type of issues commonly raised by Safety Auditors. The authors hope that the book will be read both by designers and Road Safety Auditors, and that the information contained will in some small way make a contribution to continuing reductions in road traffic accident casualties.

Figure 8.6
Radical approach to road safety promotion adopted by the Public Roads Administration in Iceland

References

1. DETR (Oct 2000). 'Highways Economic Note 1'
2. IHT (1995). 'Review of Road Safety Audit Procedures'
3. Surrey County Council Monitoring Report
4. Ministry of Transport, Denmark (1997). 'Manual of Road Safety Audit, Road Directorate'
5. TRL (1999). 'The Benefits of Road Safety Audit, paper presented at European Road Safety Conference', Malmo
6. DETR (March 2000) 'Tomorrow's Roads – Safer for Everyone'

Index

Note: Figures and Tables are indicated (in this index) by *italic page numbers*